TWO PATHS, 1 FUTURE

Ownership, Systems, and the Clarity We Need

Theodore McDuffie

First Edition

Table of Contents

Chapter 1
THE WORLD IS SHIFTING

A World in Tension

Something is changing. It is not moving slowly or quietly, but in a way that is fundamentally reshaping how the world works—how value is created, how systems operate, and how people participate within them. You can feel it, even if you cannot fully explain it yet.

There is a tension in the air between what has always worked and what is beginning to replace it; between stability and transformation; between the systems people trust and the systems that are quietly taking their place. Most people feel it, few understand it, and even fewer are prepared for what it means.

Because this is not just a shift in technology. It is a shift in how the world organizes value.

A Moment of Awareness

Across nations and communities, a quiet awareness is rising that something is no longer stable. Information moves instantly, but clarity does not. We are surrounded by noise—opinions, narratives, and constant signals—all competing for our attention. Truth blends with distortion until it becomes difficult to separate what is real from what only appears to be.

And in that confusion, something deeper begins to emerge. When noise reaches its peak and division becomes constant, people begin searching for something steady, real, and true. Beneath it all, a question begins to surface: *What can I trust*? At the same time, the global system feels strained. Institutions are questioned and foundations feel uncertain. This is not just pressure on systems—it is pressure on perception. We are in a moment that asks whether we will evolve to be more aware, intentional, and connected, or return to familiar patterns of division and conflict.

What we are witnessing is not random change. It is direction.

The Digital Mirror

As systems evolve, so does the environment surrounding us. Data is constantly being collected, stored, and studied. Every click, every search, and every interaction is a signal. We are not just using these systems; we are shaping them, training intelligence, and feeding algorithms. In return, they begin to shape us.

This raises a deeper question: Are we guiding these systems, or are they quietly guiding us? Pause for a moment and ask yourself something simple: Can you disconnect, even briefly? Because beyond all systems, something real still exists.

The Beauty We Forgot to See

Think of the air, the sun, and the rhythm of nature—the beauty of real life in its purest form. There is fresh air filling your lungs in a way no system can replicate, the warmth of sunlight touching your skin after a long winter, and the quiet, patient movement of water across a lake.

Green trees stretch toward the sky with roots deep in the ground, unmoved by trends and untouched by algorithms. We see parks filled with life: children laughing, people walking, and conversations happening in real time, not through screens. The sound of wind moving through leaves and birds calling in the distance are reminders of what life feels like when it is experienced—not filtered, optimized, or constantly interrupted.

It has always been here: accessible, available, and free. And yet, many people no longer see it. Not because it disappeared, but because our attention moved elsewhere—toward systems that demand constant engagement and environments designed to capture focus. Slowly, without realizing it, we begin to disconnect from what is real while becoming more connected to what is constructed.

This is where awareness becomes essential. You do not recognize the value of something until it is no longer within reach. Clean air matters when it is polluted. Silence matters when it is replaced with constant noise. Presence becomes powerful when distraction becomes the norm.

The real world has not lost its beauty, but humanity risks losing its ability to see it. It is about attention: where it goes, what it feeds, and what it forgets.

Take a moment. Step outside. Breathe deeply. Feel the air and notice the sky. Listen to what is not demanding your attention, but has always been there. Before systems, technology, or structures of ownership, there is life.

Within that life, there is something that cannot be replicated, digitized, or replaced.

How Change Actually Happens

Transformation rarely announces itself. It does not arrive loudly; it moves quietly beneath the routines and patterns people have learned to trust. Daily life continues—people go to work, build businesses, and raise families—and yet, beneath it all, the systems that organize value, define ownership, and shape opportunity begin to shift.

Every generation reaches a moment like this where the foundation evolves. It happens steadily, and often, by the time the shift becomes clear, it is already underway. Those who recognize it early do not just experience change. They position within it.

Where the Shift Became Visible

In 2008, a signal appeared in the form of a white paper and a new idea: Bitcoin. At the time, it seemed small and easy to ignore. But it introduced something deeper—a new way to record, verify, and transfer value securely and digitally without centralized control. For the first time, ownership itself could exist independent of traditional institutions.

It raised a question few fully understood at the time: If currency can be owned this way, what else can?

A New Ownership Conversation

Today, that question is no longer theoretical. It is being explored by institutions, governments, and innovators alike. New systems are emerging that reshape how assets are recorded and how transactions move. This is where the tokenization of real-world assets takes form.

To many, it still feels technical and abstract because most are focused on the work of everyday life. Yet, beneath all of it, there is something connecting everything: assets. A business is an asset. A building is an asset. A brand or an idea is an asset. Even communities create measurable value.

The question is no longer whether value exists. The question is: **Who owns it?**

Why This Moment Matters

Most people have not yet been invited into this conversation, and that is exactly why this moment matters. While many are still trying to name what they feel, the system has already begun to move. We are early in the formation of these new structures.

Being early does not remove risk; it exposes it. But early is also the only moment where awareness can still become an advantage. There is still time to observe without pressure and to understand before participation becomes unavoidable.

This is about seeing how value is created and how ownership is assigned. Once you see the structure, you realize something most people miss: the outcome was never random. What once felt distant comes into focus. What once felt complex reveals its design.

With that clarity, something shifts. Not just belief, but position.

The First Fracture

Awareness does not comfort you; it confronts you. What felt stable begins to show strain, and what once felt like a system you lived inside begins to feel like a system under structural pressure. This pressure does not ask for attention—it eventually demands it. When that moment comes, it will not begin with an explanation. It will begin with impact.

This shift is being driven by imbalances forming for years. Every system, no matter how strong it appears, has a breaking point. The question is no longer whether pressure exists, but how much is already there.

The signals are no longer quiet. Markets are shifting, technologies are accelerating, and trust is being tested. This is not just disruption; it is transition. When pressure builds inside a system, it does not remain the same.

It adapts, it restructures, or it breaks.

Chapter 2
A GLOBAL SYSTEM UNDER STRAIN

The Fragility of the Pattern

To understand this moment, you have to look beyond the headlines and see the pattern. Tension is rising globally as alliances shift and long-standing assumptions are challenged. Critical infrastructure, once taken for granted, is now being recognized for what it truly is: fragile.

Take energy, for example. A single geographic point like the Strait of Hormuz handles a significant portion of the world's oil supply. Any disruption there does not stay regional; it becomes global. Prices move, markets react, and entire economies feel the impact. This is what pressure looks like when the system is tightly connected and heavily dependent on stability that cannot be guaranteed.

The Currency Question

At the same time, another shift is unfolding. For decades, the United States Dollar has served as the foundation of the global financial system—trade, reserves, and settlements were all anchored in one dominant currency.

Now, countries are exploring alternatives through new agreements and settlement systems. This process, often described as "De-dollarization," is not a sudden collapse, but a gradual repositioning away from a single point of control toward a more distributed system. When the foundation of global finance begins to shift, everything built on top of it feels the pressure.

The Weight of Debt

Then there is the national debt, a number that continues to grow quietly and relentlessly, now measured in the tens of trillions. This is a system sustained by borrowing and maintained by the belief that it will continue to function.

But debt is more than a number; it is a signal of how systems are structured and how future obligations are being created faster than they are being resolved. Over time, that weight compounds, not always visibly, but structurally.

Technology and the End of Labor

While this financial pressure builds, technology accelerates. Artificial intelligence, automation, and robotics are scaling at an unprecedented rate. For the first time in history, technology is not just enhancing labor—it is replacing it.

Machines produce, systems optimize, and algorithms decide. This leads to a deeper question: What happens when human labor is no longer the primary way people participate in the economy?

In past economic shifts, new jobs eventually replaced old ones. But this transformation may not follow that pattern. There may not be enough new roles to absorb what is lost. If participation remains tied to labor while labor becomes less necessary, the gap widens exponentially. The systems producing value continue to grow, while fewer people remain connected to that growth.

This is how economic inequality accelerates: not just through effort, but through design.

What Pressure Reveals

Pressure removes the surface and forces reality into view. What was once hidden becomes visible; what was once assumed is now questioned. Because every system is designed and every outcome has a pattern behind it, we realize that this is not random. It never was.

From this pressure, two paths begin to form: one where control increases and power concentrates, and another where participation expands because people understand how to engage with value differently. The decisions being made now are shaping which path becomes our reality.

The Shift from Labor to Ownership

For generations, most people participated through work, exchanging time and skill for income. That model built modern life, but it also preserved a divide: most people earned, while a smaller group owned.

Ownership determines long-term participation in value. A worker earns once; an owner participates continuously. That difference is becoming more important than ever. If systems can produce more value with less human labor, then participation must evolve beyond labor alone.

The Widening Divide

Global tension, energy instability, currency shifts, rising debt, and technological disruption are not separate issues. They are signals of a system under pressure. When a system is under pressure, it doesn't just change at the top; it changes how every individual participates within it.

Beneath all of this, a deeper divide is forming. It is not just a gap in income or opportunity, but a gap in understanding, access, and ownership. It is a divide between those who are positioned to participate in what is being built, and those who will experience it from the outside looking in.

This gap did not begin today. It has been forming over generations, quietly, until now. Before we can understand what is coming, we have to understand the gap between effort and long-term value. Once you see that gap clearly, the entire conversation changes.

Chapter 3
THE OWNERSHIP GAP

The sun was dropping behind the stadium lights as practice came to an end. The sound of helmets colliding faded into the evening air, replaced by the chatter of players pulling off their gear. They walked off the field in groups—laughing, talking, decompressing from the work they had just put in. Some replayed specific moments from practice, while others talked about school, relationships, and the life that existed beyond the field.

Others were already locked in on the next game. A few believed football could take them to college, and one or two even believed they might reach the professional level. Many were simply hoping the game would open doors that life had not always made easy.

I stood near the sideline, watching them walk away. As a coach, you don't just see athletes; you see potential, pressure, and promise. But if you're paying attention, you also see something else—something most people don't talk about. You begin to see how uneven the starting line really is.

While they all wore the same uniform, they were not all playing the same game.

Not when it came to opportunity. Not when it came to access. And definitely not when it came to ownership. Standing on that field, I realized something clearly: The gap was never just about talent. It was about what happens after the game. Because talent can open doors—but it does not determine what's waiting on the other side.

I've seen young men with discipline, work ethic, and heart give everything they had to the game. They endured early mornings, late nights, pain, and sacrifice. They learned structure, execution, and how to perform under pressure. But very few were ever taught how to own. Not just to own their effort—but to own assets, decisions, and opportunities. They weren't taught to own a future beyond the systems they were trained to operate within.

And that is where the gap begins.

What the Ownership Gap Really Is

The Ownership Gap is the difference between participating in systems and owning within them. It is the distance between those who are connected to how value is created, structured, and multiplied—and those who are not. It is not just about money; it is about understanding and access. It is about who knows how value is created, how it moves, and how it grows over time.

At its core, the Ownership Gap is the difference between participating in a system and positioning yourself within it. It is the shift from earning to building, and from working to owning.

For generations, many people were taught how to function inside systems, but not how to benefit from them. They were taught how to earn, but not how to structure ownership; how to survive, but not how to build long-term value. Over time, that difference compounds quietly until it becomes impossible to ignore. You begin to see it everywhere: in who owns the buildings versus who rents inside them; in who controls capital versus who depends on access to it; in who benefits from growth versus who is simply affected by it.

Now, as technology reshapes how value moves and new forms of ownership emerge, this gap will not remain the same. It will do one of two things: It will close through awareness, access, and intentional participation—or it will widen faster than ever before.

The future will not automatically include everyone. It will include those who understand it and prepare for it.

The Lesson Beneath the Game

That evening, as the field emptied and the lights came on, a deeper realization stayed with me. We were teaching them how to compete, how to push beyond limits, and how to stay disciplined toward a goal. But we were not teaching them how ownership works. And the truth is—for most of my life—no one had taught me either.

Like many people, I grew up believing that hard work was the sole foundation of opportunity. You work hard in school, in sports, and in life. Those lessons matter; they build discipline, resilience, and character. But over time, I began to see something deeper: Hard work alone does not explain how opportunity is created or sustained.

Growing up in Buffalo and later coaching in communities like it, I was surrounded by people who understood effort. People who showed up every day and pushed through real challenges. Junior college football made that reality even clearer. Many players had no scholarships, no meal plans, and no guarantees—just practice, class, and the pursuit of opportunity, often with very little support beyond sheer determination.

And yet, despite all that effort, something didn't fully add up. The people I knew were working incredibly hard, but the individuals and institutions building long-term wealth had something different: **Ownership.**

Labor vs. Ownership

For most of modern history, people have participated in the economy through labor. They trade time, energy, and skill for income. There is nothing wrong with that; every system depends on labor. But labor and ownership function in fundamentally different ways.

Labor produces income. Ownership produces assets.

Assets have the ability to do something labor alone cannot: they can grow, they can generate income over time, they can be transferred, and they can shape opportunity across generations. A business, a building, a piece of land, a brand, or intellectual property—all of these are assets. This distinction is simple, but it is not widely taught. Most people learn how to work; very few are taught how assets work.

Ownership changes your relationship to value. Instead of being paid for your time, you become connected to growth. Instead of starting over each day, you begin to build something that continues. You begin to develop sources of value that can expand beyond your direct effort. It shifts you from participating to positioning.

Once You See It, You Can't Unsee It

Once I recognized the pattern, I saw it everywhere: in classrooms where financial systems are rarely explained; in athletes who understand competition but not contracts; in entrepreneurs building businesses without understanding how they become scalable assets; and in communities creating enormous value while ownership remains largely invisible.

The pattern became undeniable. People are working, creating, and contributing, but very few are taught how the systems that organize ownership actually function.

Across cities like Buffalo and countless others, the effort is undeniable. Small business owners open their doors before sunrise, nonprofit leaders stretch resources to serve others, and artists shape our culture. The value is everywhere. But ownership is not always visible. If value is being created everywhere, why isn't ownership distributed the same way?

The answer is not random. The Ownership Gap exists because of how systems have been designed. Over time, systems were built to prioritize efficiency, centralize control, and reward those who understood how to structure ownership—not necessarily those who created the most value.

Access to ownership has never been evenly distributed. It has been shaped by exposure (who is taught how systems work), access (who has entry points), and positioning (who is placed close enough to value to participate). For many, those pathways were never clearly shown.

The System Beneath the Surface

Gaps do not appear by accident. They are built over time through design and through what is never explained. What systems prioritize, they expand; what they ignore remains invisible. For a long time, broad participation in ownership was not the priority, so value continued to be created while ownership remained concentrated.

As systems evolve again—as technology reshapes how value moves and new models of ownership emerge—this gap does not disappear. It follows the system. If the structure is not understood, it will be recreated, even inside new technologies and new opportunities. That is why awareness alone is not enough. You can see the system and still participate in it the exact same way.

Understanding structure is what allows it to change.

Once you recognize that the gap is structural, the conversation shifts. Surface questions lose their power. It is no longer enough to ask who has ownership; that only describes the outcome. The real question is deeper: **How was ownership designed this way?** What forces shaped it? What decisions reinforced it?

Ownership is not fixed. It never has been. It has always been built, adjusted, and redefined across moments of change. What exists today is not permanent; it is the result of evolution. And because evolution does not stop, we must step back to study the pattern.

Only by understanding where ownership has been can you recognize where it is going—and more importantly, where you stand as it moves. The next version of ownership will not arrive as an announcement. It will emerge quietly through the same forces that have always shaped it.

By the time it becomes obvious, it will already be in place.

Chapter 4
THE EVOLUTION OF OWNERSHIP

From Land to Code

Why Ownership Must Be Understood

To truly comprehend the trajectory of our economic future, we must pause to consider a fundamental truth that most people take for granted: ownership has never been a static concept. It is not a fixed monument, nor is it a permanent set of rules carved into stone. Throughout human history, ownership has been a dynamic, living force, constantly shifting, reshaping, and reorganizing itself. It is continuously redefined by the arrival of new technologies, the reorganization of power structures, and the evolution of global systems. Every time the world undergoes a major societal shift, the architecture of ownership changes with it. This transformation does not happen gradually at the margins; it happens structurally at the very foundation of how we define value.

The First Foundation — Land

At the dawn of organized human economies, the concept of ownership was remarkably simple, tangible, and absolute. Land was the ultimate measure of power. To own the soil was to control the means of production, the distribution of vital resources, and survival itself. Agricultural societies were built entirely upon this singular, unyielding principle. In this era, the social structure was clear-cut: those who held title to the land owned the entire system, while those who did not were relegated to working within it. This division created the earliest, most visible iteration of the Ownership Gap. There was no ambiguity or complexity here; ownership was a physical reality that directly dictated an individual's position in the social hierarchy. If you did not own the earth beneath your feet, your labor belonged to someone else.

The Industrial Shift — Production and Scale

Eventually, the world underwent a seismic transformation. Steam engines hissed, iron machinery roared to life, and factories rose like monoliths across the landscape. The Industrial Revolution fundamentally redefined the nature of human existence, and with it, the mechanics of ownership evolved. Value was no longer bound solely to the acreage of fertile land. Instead, it migrated into the machinery of production: factories, assembly lines, energy grids, and global supply chains. Those who acquired ownership of these production systems held a new kind of power. They did not just control raw localized resources; they controlled global scale. Although this industrial shift expanded the labor market, created millions of new jobs, and drove the rapid growth of modern cities, the underlying pattern remained stubbornly familiar. The vast majority of the population exchanged their physical labor for a wage, while a select group owned the infrastructure. The system had evolved to look entirely different, but the fundamental structure of the gap had only expanded.

The Financial Era — Capital as Control

As global economies matured and grew more sophisticated, ownership underwent yet another profound evolution. It began to shed its heavy, physical skin, becoming increasingly abstract, modular, and legalistic.

Liquid capital gradually replaced land and heavy machinery as the dominant force of economic control. Ownership migrated into paper assets, legal frameworks, corporate entities, equity shares, stock markets, and complex financial instruments. Value was no longer tied directly to what you could physically touch or see; it was tied to what you could legally structure and negotiate. This era birthed layered, highly complex systems of ownership that were largely invisible to the everyday workers operating within them. Once again, those who understood this new, abstract architecture participated in the economy at a highly leveraged level, while those who did not were left to trade their finite time for a depreciating currency.

The Digital Era — Platforms and Data

Then came the dawn of the internet—a hyper-connected, digital ecosystem that introduced an entirely new layer of value to the human experience. Physical spaces were rapidly replaced by digital platforms. Information became the primary raw material, data became the most valuable asset class on earth, and human attention became the ultimate global currency. In this landscape, ownership evolved once again. The individuals and corporations that built and owned the dominant digital platforms suddenly controlled entire global ecosystems. They did not merely participate in the creation of value; they directed and governed it at an unprecedented, borderless scale. Yet, for the average person, this digital layer of ownership remained entirely invisible. The systems were designed to be incredibly frictionless and easy to use, but their underlying ownership structures were deeply complex and highly concentrated. We became consumers of platforms we did not own, training algorithms that we could not control.

The Signal — A New Foundation Emerges

In the midst of global financial instability in 2008, a quiet but revolutionary signal appeared in the form of a white paper. It introduced a new digital asset: Bitcoin. To the casual observer, it looked like nothing more than an experimental, volatile alternative currency. But beneath the surface of this technology lay something far more significant: a fundamental breakthrough in how human beings structure trust. For centuries, the transfer and validation of ownership required trusted intermediaries—banks, governments, registries, and centralized institutions. These entities held the master ledgers, verified transactions, and controlled systemic access. Bitcoin introduced a radical alternative: a cryptographically secured system where value could be recorded, verified, and transferred globally without the need for centralized control. It was transparent, mathematically rule-bound, and distributed. This was not just a clever technological innovation; it was a profound structural shift. For the very first time in human history, authentic ownership could exist digitally, globally, and completely independently of traditional centralized authorities.

From Currency to Infrastructure

What began as an experiment in digital currency quickly revealed a much deeper, more expansive systemic possibility. If the ownership of money could be digitized, secured, and decentralized in this manner, then what else could be? The answer was simple and ultimate: anything of value. Land, commercial real estate, private businesses, intellectual property, creative masterworks, and municipal infrastructure could all be recorded on this new digital foundation. This realization shifted the global conversation from cryptocurrency to foundational architecture. Out of this technological bedrock emerged blockchain technology, self-executing smart contracts, and the concept of tokenization. These were not passing trends or speculative bubbles; they were the opening salvos of an entirely new ledger system for human civilization.

Tokenization — The Next Evolution

Tokenization represents a monumental leap forward in the history of property rights. It is the technological process of taking real-world, illiquid assets and representing their ownership digitally on an immutable ledger. This allows ownership to become highly divisible, frictionlessly transferable, globally accessible, and digitally programmable. This is no longer a theoretical exercise confined to academic circles; it is actively being engineered, tested, and integrated by major financial institutions, sovereign governments, and global enterprise systems. Tokenization solves a historical problem that every previous economic system struggled to overcome: it connects value directly to fractional participation. It provides a technical mechanism to democratize access to high-value assets. Yet, like every major paradigm shift that preceded it, this technology will not distribute its opportunities evenly on its own. The infrastructure is neutral, but the benefits will inevitably flow to those who understand its design.

The Pattern That Never Changes

Every distinct economic era in human history follows an identical, predictable cycle. A new system of value and ownership emerges. At first, it is widely misunderstood, dismissed by experts, and overlooked by the masses.

Slowly, however, it gathers momentum, proves its utility, and quietly integrates itself into the background of daily life until it becomes the dominant infrastructure of society. Those who take the time to understand the system in its early, unpolished phases are able to position themselves within it. This advantage does not exist because these early adopters are inherently smarter or more talented than their peers; it exists because they possess a rare currency: timely awareness. Awareness is the raw material that shapes structural position.

Where We Are Now

Today, we are not standing on the threshold of this economic transition; we are already living inside of it. The very definition of ownership is undergoing its next great transformation, moving from the physical to the industrial, from the industrial to the financial, from the financial to the digital, and now, to the programmable. Ownership is becoming dynamic, liquid, and capable of moving at the absolute speed of light. This changes the fundamental question of human economic survival.

The evolution of ownership has never been an abstract or academic debate; it has always revealed itself through real-world moments. These are the quiet, historical inflection points where the ground beneath our feet begins to shift, reorganizing society in ways that people cannot immediately put into words, but can deeply feel. If you study these moments of transition closely, you will find that history leaves clear clues. These clues are not found in theoretical economics papers, but in real physical places where the future chose to show up early—places where the invisible structures of tomorrow first became visible, allowing those present to experience the future before the rest of the world even had the language to describe it. One of those historic, foundational places was Buffalo, New York. In 1901, the darkness of the physical world did not simply fade away; it was instantly conquered by a brilliant, blinding light. It was a signal of a new age, guided by the legendary mind of Nikola Tesla—a man who did not merely study the mechanics of electricity, but who deeply understood the cosmic patterns of frequency, energy, and vibration, often symbolized by the mathematical relationship of 3, 6, and 9. What thc crowds saw as mere physical illumination was, in reality, a profound demonstration that a new way of organizing human life and energy had arrived.

Chapter 5
BUFFALO: THE FIRST LIGHT

A Moment That Changed Everything

1950 Delaware Avenue – Buffalo, New York

This historic building stands quietly as one of only two remaining structures from the grand Pan-American Exposition of 1901—a spectacular, brief moment in time that fundamentally redirected the current of human history. Within this space, a stunning mural pays tribute to the resilient spirit and heritage of the people of Buffalo. Commissioned as a gift of love by Carol M. Coppola and brought to life by the hand of artist Tim Martin, this artwork honors a profound turning point that did not just change the fate of a regional city, but altered the trajectory of the entire modern world. In the autumn of 1901, Buffalo became the very first city in the United States to be illuminated at scale by high-voltage alternating current electricity, made possible by the pioneering genius of Nikola Tesla. What the awe-struck onlookers perceived as mere light shining through the dark was actually something far more profound. It was the birth of modern electrical infrastructure. It was the materialization of a new system of energy distribution.

Dimension	Historical Parallel: Tesla's AC Grid (1901)	Modern Parallel: Distributed Ledger / Tokenization
The Visible Layer	**Physical Light:** Illuminated streets, shining incandescent bulbs, and immediate visual spectacle.	**Token Price & Hype:** Volatile exchange rates, market symbols, and immediate speculative activity.
The Unseen Layer	**Invisible Grid of Power:** A massive network of copper lines, step-down transformers, and physical power generators.	**The Trust Engine:** Decentralized ledgers, smart contracts, and secure mathematical infrastructure.

What This Really Represents

At the turn of the century, the average person standing in the glow of the Exposition could not fully articulate or comprehend the monumental nature of what they were witnessing. They saw beautiful light; they felt an exciting, emotional rush of human progress. But directly beneath that beautiful spectacle, an entirely new layer of physical and economic reality was being constructed. A massive, invisible systemic web of wires, generators, transformers, and grids was being laid down—a system that would go on to permanently dictate how modern cities would function, how heavy industries would scale, how global economies would expand, and how everyday human beings would live their lives. This is the precise template of true structural transformation. It does not arrive with a loud, warning announcement; it integrates itself quietly, building the invisible systems that redefine the parameters of what is possible for humanity.

The Connection to Now

This historical moment in Buffalo is not a dusty chapter of the past; it is a recurring blueprint of societal evolution. Every major paradigm shift in human civilization begins in this exact sequence: a revolutionary new system emerges; a small, focused group of individuals understands its profound structural implications early; and the vast majority of the public experiences the surface-level effects of the change without ever truly seeing or understanding the machinery operating behind the curtain. Alternating current electricity was never just about replacing oil lamps with lightbulbs, just as today's decentralized ledger technologies are not merely about digital assets or financial speculation. These technologies are about foundational structure, ownership, and systemic participation.

From Light to Systems

Buffalo was rightfully crowned the "City of Light," yet the true genius of that historical moment was not the visual illumination of the streets. The real story belonged to those who understood the massive, industrial power generation systems being quietly assembled beneath the surface. The individuals who recognize the value of systemic infrastructure in its infancy do not merely live through the transition; they position themselves strategically within its architecture.

Gratitude and Recognition

We owe a deep debt of gratitude and recognition to Carol M. Coppola, Tim Martin, and the immortal legacy of Nikola Tesla for their permanent, cascading impact on the city of Buffalo and the broader world. The future rarely announces its arrival with a formal invitation. Instead, it is built quietly in moments that look like simple technological progress, but are actually the cornerstones of entirely new societal systems. History demonstrates that these rare inflection points do not belong exclusively to one geographic location, one static era, or one generation of thinkers. They echo across time. They return in cycles. They wait patiently to be recognized by those with the eyes to see them.

As the modern world hurtles forward through its next great transition—navigating the complexities of programmable digital assets, automated systems, and decentralized networks—the ultimate question shifts from the global to the local, from the abstract to the deeply personal. Could a city like Buffalo, which once acted as the literal spark that illuminated an entire nation, rise to become the epicenter of a new kind of light? This time, the challenge is not just about lighting physical streets, but about illuminating human understanding. It is not merely about powering industrial machines, but about helping ordinary people clearly see and navigate the economic networks that govern their daily lives. What if the next great historical shift is not defined by technology alone, but by a massive, grassroots awakening of systemic awareness? This is the kind of awareness that reconnects local communities directly to the tangible value they generate every single day. This is the clarity that aligns physical participation with structural ownership, finally working to close the painful, generational gap that has divided our society for far too long. This bridge cannot be built by mere hope or random chance; it must be constructed through structural design, educational clarity, and a community of people who are willing to see the world differently and build accordingly. Sometimes, the path to the future does not require us to search for something entirely new. Sometimes, the light we need is simply reignited from a place that has already shown the world what is possible. These moments of historical triumph serve as powerful reminders that what once felt completely impossible can become our everyday reality the moment we choose to look past the noise, perceive the underlying pattern, and step forward before the rest of the world understands what is unfolding. True transformation does not begin with the adoption of a new tool; it begins with a fundamental shift in human perspective.

Chapter 6
THE OWNERSHIP MINDSET

From Awareness to Alignment

Where the Shift Becomes Real

Up to this point in our journey, we have pulled back the curtain to examine the global system. We have mapped the immense macro pressures building within it, analyzed the historic depth of the Ownership Gap, and traced the long evolutionary line of how property rights have transitioned from physical land to digital code. But none of this systemic knowledge changes a single detail of your life unless it sparks a fundamental transformation within you. External systems do not automatically elevate people. The raw accumulation of information does not liberate individuals, and passive awareness alone possesses no power to change your life. True transformation requires alignment. This is the exact point where our conversation must shift from the grand movements occurring on the global stage to the quiet, internal evolution occurring within your own mind.

The Identity Barrier Model

The journey toward true alignment is not an unobstructed pathway. To understand why some cross the divide while others remain on the far shore, we must examine the cognitive gatekeeping process of the human mind.

Systemic Phase	**Cognitive State**	**Core Developmental Dynamics**
Phase 1: External Shift	Passive Awareness	The physical or technological arrival of new economic systems, ledgers, and programmable assets.
The Core Barrier	**Cognitive Bottleneck**	The internal identity crisis and subconscious gatekeeping question: "*Am I actually allowed to own assets, or am I just designed to work for those who do?*"
Phase 2: Real Alignment	Active Ownership	The internalization of the ownership identity; stepping out of the passive loop to become a systemic builder.

The Difference That Separates Outcomes

It is a fascinating, repeating phenomenon of human behavior: two individuals can observe the exact same macroeconomic shift, receive the identical financial information, and stand before the very same structural opportunity, yet walk away with completely opposite life outcomes. This disparity does not exist because one individual possessed superior systemic access, nor does it exist because they had access to more capital. The difference lies entirely in an internal decision—a choice of how they see themselves in relation to the world around them. This internal divide is the presence or absence of the ownership mindset.

What the Ownership Mindset Actually Is

To understand the ownership mindset, you must first understand what it is not. It is not merely the possession of material assets, nor does it begin with a large bank account. In its earliest phase, this mindset is a matter of pure cognitive perception. It is the specific mental lens through which you perceive value, interpret global systems, and position your life in relation to the infrastructure being built around you. It is a profound, conscious cognitive shift that moves you along a clear spectrum:

- **From a passive consumer of platforms** to an active student of their structural design.
- **From a reactive participant in market cycles** to a proactive positioner of wealth.
- **From participating solely in a labor force** to structuring ownership within the value loop.

True ownership does not begin with what you hold in your hands; it begins with how you choose to see with your eyes.

Why Most People Never Move

While systemic awareness is the essential catalyst that creates the opportunity for a personal shift, it provides no guarantee that the shift will actually occur. It is entirely possible for an individual to see a system's flaws with crystal clarity and yet choose to remain frozen in their current position. They can possess a deep, intellectual understanding of economic changes and yet never alter the way they interact with those systems. They can spot a life-changing opportunity on the horizon and still fail to take a single step toward it. This inertia occurs because intellectual understanding alone is powerless to override years of deep social conditioning, and new information cannot easily replace a deeply ingrained personal identity.

The Real Barrier

The invisible wall that prevents most people from stepping into ownership is not the obstacle they think it is. It is rarely a lack of information, an absence of starting capital, or a complete lack of opportunity. The ultimate, unspoken barrier is identity. It is the unconscious belief system that dictates how you see yourself in relation to wealth and ownership. It is the quiet voice that asks: *Am I the kind of person who is allowed to own things, or am I simply designed to work for those who do?* If the concept of ownership has never been integrated into your core identity, you will unconsciously sabotage your efforts and refuse to move toward it, even when the structural shifts of the world are aligned perfectly in your favor.

The First Shift — From Passive to Intentional

The vast majority of people are trained to move through life in a completely passive relationship to the systems around them. They follow the pre-paved path of least resistance that was laid out for them by previous generations: show up to work, earn a wage, spend what remains, and repeat the cycle. While this traditional path is honorable and capable of building disciplined character, it was never designed to produce structural ownership for the average participant. Developing an ownership mindset requires a deliberate, courageous departure from this passive loop. It is the choice to transition from merely reacting to systems to actively observing them; to stop just moving through the economy and start analyzing its design. It requires you to ask entirely different questions:

- *What is the underlying system I am currently operating inside of?*
- *Who owns the infrastructure of this system?*
- *How is value actually generated here, and where does it flow?*

Changing the quality of your questions is the only way to change what you see, and what you see directly dictates how you choose to move.

The Second Shift — From Earning to Building

For most of modern history, human beings have been conditioned to believe that their personal value is directly and exclusively tied to their physical or mental effort. We are trained to think in a linear equation: time in equals money out. However, the ownership mindset introduces a completely different relationship with value. It moves you past the limits of earning and introduces you to the power of building. A builder does not think in terms of hourly wages or immediate payouts; they think in terms of systemic scale and longevity. They ask:

- *What can I construct today that will continue to exist and produce value without my physical presence tomorrow?*
- *How can I connect my current daily efforts to a structure of long-term asset growth?*

This is the precise entry point where assets transition from a luxury of the wealthy to an essential survival structure for everyone.

The Third Shift — From Consumption to Structure

Modern societal systems are brilliantly, ruthlessly engineered for consumption. They are designed to keep your mind constantly engaged, your emotions continuously reacting, and your attention trapped inside an endless loop of spending. The ownership mindset requires a completely different, counter-cultural posture. It is the discipline to look past the surface of products and platforms to study how they are structured. Once you understand the underlying architecture of a system, you gain immediate, powerful economic leverage. You begin to trace where value originates, how it moves through networks, and who ultimately captures it at the end of the line. Once this structural clarity is achieved, it becomes impossible for you to participate blindly in the economy again.

The Fourth Shift — From Fear to Clarity

Periods of rapid, global systemic change naturally produce high levels of social uncertainty, which in turn breeds widespread fear. When people are afraid, they instinctively react, making emotional decisions to protect what they have. But reaction is the exact opposite of strategic positioning. The ownership mindset refuses to let fear drive its decisions. It is grounded in a calm, analytical clarity that allows you to pause, step back, and observe the landscape before taking action. While fear asks anxious questions like "*What if I get left behind or miss the trend?*" clarity asks steady, powerful questions: "*What is actually happening beneath the noise? What is structurally real, and how does it align with where the world is going?*" This single cognitive difference is what separates those who are swept away by change from those who ride its wave.

The Fifth Shift — From External Systems to Internal Alignment

At the absolute deepest level, the journey of ownership is not an external struggle against the world; it is an internal alignment with truth. You can be completely surrounded by life-changing opportunities and yet remain entirely blind to them if your internal state is out of alignment. You can be handed absolute systemic access on a silver platter and still fail to take action if you do not believe you belong in the room. Access without internal alignment changes absolutely nothing, and structural opportunity without conscious awareness will always go unused.

The Missing Layer — Ownership Literacy

Yet, even if you successfully make all five of these cognitive shifts, there is still a critical piece of the puzzle missing. Changing your mindset alters how you see the world, but it does not automatically teach you the practical, technical rules of how to navigate it. This is the exact point where most aspiring builders stall out. They train themselves to think differently, but they continue to operate using the exact same outdated tools.

A New Literacy

Every major structural transformation in human history births entirely new systems of operations. Those who take the time to learn the technical literacy of these new systems in their infancy are granted a fundamentally different level of economic participation. This reality points to a simple, undeniable truth: the most critical literacy of the coming era is not traditional financial literacy. It is ownership literacy. This is the practical ability to understand:

- How digital and physical assets operate in a connected world.

- How ownership is legally and technologically structured.
- How value moves across borderless, decentralized networks.

This literacy is not taught to induce fear; it is cultivated to provide absolute economic clarity.

Why Ownership Literacy Matters

The historical record is incredibly consistent: moments of technological and systemic transition create unprecedented windows of opportunity, but that opportunity is never distributed evenly at the start. The Industrial Revolution birthed factories, the digital age birthed platforms, and our current era is birthing a completely programmable financial infrastructure. We are entering a phase of human history centered entirely on how property rights are represented, structured, and accessed globally.

Mindset Opens the Door — Literacy Shows the Path

The ownership mindset is the force that opens the door of opportunity, but ownership literacy is the map that shows you exactly how to walk through it. Mindset alters your perception, while literacy directs your execution. To possess one without the other is to remain fundamentally incomplete.

From Awareness to Execution

This is where the vast majority of people fall behind in times of change. They achieve a level of intellectual awareness, but they never commit to acquiring the practical literacy required to execute. They watch the world shift from the sidelines, remaining passive observers of systems they have every right to participate in and own.

Ownership Is a Decision

Ultimately, this journey stops being about the accumulation of information and becomes an act of pure, decisive will. It is the decision to see the world clearly, to ask difficult questions, and to position your life with absolute intentionality. Ownership is not a privilege you wait to be granted by an authority; it is a structural position you choose to step into. It begins in the quiet landscape of how you think, manifests in the daily actions of how you move, and culminates in the lasting value of what you choose to build and own.

CLOSING — ALIGNMENT

Ownership is not a subject you simply study and memorize; it is a state of being you must align your entire life with. Mindset shapes your perception, literacy directs your execution, and alignment determines the ultimate legacy of what you build. But even this is not the final, deepest layer of the structure. Beneath mindset, beneath literacy, and beneath your daily decisions, there are invisible laws and recurring patterns that shape human perception, guide public behavior, and dictate economic outcomes without the masses ever being aware of their existence. What you cannot see is still moving you.

Chapter 7
THE HIDDEN LAWS OF AWARENESS

What You Don't See... Is Still Moving You

Where Awareness Stops Being Optional

There inevitably comes a specific point in every societal transition where ignorance ceases to be a neutral state of existence. There is a moment where not knowing the rules of the game is no longer a harmless oversight, but a form of active economic self-sabotage. The physical and economic systems of the world do not pause their evolution to wait for those who do not understand them. The world moves forward ruthlessly, with or without your awareness, with or without your explicit consent, and with or without your participation.

The Illusion of Control

The average person moves through their daily routine under the comforting illusion that they are making completely independent, rational decisions. What they fail to perceive is that they are merely reacting to deep, systemic patterns they have never been trained to recognize. These are patterns that engineered their desires, patterns that systematically cultivated their fears, and patterns that placed invisible limits on their potential. When these structural patterns go unseen, they do not lose their power; instead, they take complete, silent control of your trajectory.

Why This Knowledge Was Never Loud

There is a profound, historical reason why this level of systemic understanding has always been communicated in quiet, guarded spaces. It is not because the truth is a highly classified secret. Rather, it is because these foundational principles are easily misunderstood when approached by an unaligned mind, and completely ignored by those who are not yet ready to take responsibility for their own lives. Once you choose to see these structural patterns, you lose the luxury of claiming ignorance. You become fully responsible for how you choose to move.

These Are Not Ideas

The principles we are about to explore are not speculative ideas, philosophical theories, or religious beliefs that you can choose to accept or reject based on personal preference. They are fundamental, immutable laws of systemic architecture. They operate continuously, whether you acknowledge their existence or not. Right now, at this very moment, these exact laws are being applied at an unprecedented global scale. They are engineered into our digital platforms, woven into our financial networks, and built into the very structures of ownership that society is transitioning toward.

Universal Hermetic Principle	Modern Systemic Manifestation	Real-World Economic Impact
1. The Principle of Mentalism	Algorithmic Engineering & Cognitive Capture	Attention capture redirecting human desire and economic focus long before financial surrender happens.
2. The Principle of Correspondence	Micro/Macro Structural Replication	Global wealth and control dynamics replicating within local community wealth disparities.
3. The Principle of Vibration	High-Velocity Tech Cycles & Market Volatility	The elimination of stationary safety; you are either accelerating forward or being swept backward.
4. The Principle of Polarity	Algorithmic Outrage & Polarization	Media systems amplifying emotional extremes to keep populations in reactive loops and out of strategic focus.
5. The Principle of Rhythm	Macroeconomic Cycles (Boom, Bust & Reset)	Periodic financial shifts that wipe out unprepared market participants while rewarding prepared strategic capital.
6. The Principle of Cause and Effect	Hidden Financial Architecture & Legislation	Engineered regulatory and technical frameworks dictating visible asset concentration outcomes.
7. The Principle of Balance	Dynamic Equilibrium of Structure and Flow	The absolute necessity of balancing asset security (structure) with active liquidity (flow) to remain autonomous.

1. The Principle of Mentalism — Where Everything Begins

The law of mentalism states that everything in the universe begins in the unseen realm of the mind. Before a behavior can manifest, there must first be a thought; before an action can be taken, there must first be a perception. This means that long before people lose their economic position in the physical world, they have already surrendered their cognitive position in the realm of thought. In a hyper-connected world where your finite attention is constantly being harvested, redirected, and algorithmically engineered by multi-billion-dollar platforms, an urgent question arises: *Are you actively thinking your own thoughts, or is your mind being quietly thought for?*

2. The Principle of Correspondence — The Pattern Repeats

The law of correspondence is famously summarized by the ancient maxim: "As *above, so below; as within, so without.*" There is no true separation in the structures of the universe. The identical systemic architectures that shape macro global systems repeat themselves at the micro level of our daily lives. The very same patterns of control and concentration that govern international economies manifest within the local structures of our immediate communities. If you are blind to the subtle operations of these patterns in the small, everyday interactions right in front of you, you will be completely incapable of recognizing them when they scale globally around you.

3. The Principle of Vibration — Nothing Is Standing Still

The law of vibration dictates that absolutely nothing in existence remains completely static. Everything is in a constant state of motion, shifting, evolving, and vibrating at varying frequencies. In the context of our global economy, this means that you are either actively increasing your level of systemic awareness and moving forward, or you are being swept backward by economic forces you do not understand. There is no such thing as a safe, neutral ground. There is no pause button, and there is no secure place to simply "wait and see." While you are frozen in hesitation, the digital and financial systems of the world are accelerating their evolution.

4. The Principle of Polarity — Control Through Extremes

The law of polarity states that everything in existence has an opposite, and that opposites are merely different degrees of the exact same spectrum. In modern systems of social control, these polar extremes are deliberately amplified to keep the public in a state of perpetual distraction. Public fear is intentionally intensified, cultural divisions are carefully engineered, and artificial scarcity is structurally reinforced. When people are pushed into these emotional extremes, they lose their capacity for objective observation. They stop thinking strategically and begin reacting impulsively. The moment your life is driven by reaction, you have surrendered your position to the system.

5. The Principle of Rhythm — Cycles Are Not Random

The law of rhythm teaches us that everything moves in predictable, repeating cycles. There is a natural tide to all things: expansion is followed by contraction, growth is followed by recessions, and periods of stabilization are followed by major structural resets. This is the natural, repeating heartbeat of global economies and technological systems. Those who do not understand this fundamental law of rhythm mistake these systemic cycles for pure, unpredictable chaos. They panic and sell their assets at the exact moment they should be calmly preparing, and they retreat into safety at the precise moment they should be positioning themselves for growth. By the time they realize the cycle has turned, the opportunity has already passed.

6. The Principle of Cause and Effect — Nothing Is Accidental

The law of cause and effect declares that every single occurrence in the universe has a specific, identifiable cause, and that chance is merely a name for a law not yet recognized. Nothing you see occurring on the global stage is a random accident. Every financial system has an intentional design, every legal structure has a specific purpose, and every economic outcome has a direct cause—even when that cause is kept completely hidden from public view. If you do not take the time to study and understand the underlying causes of economic shifts, you are condemned to live forever as a passive victim of their effects, constantly reacting to outcomes you were never taught how to influence.

7. The Principle of Balance — Structure and Flow

The law of balance dictates that all true creation and sustained progress require a perfect equilibrium between structure and flow. Structure without flow becomes stagnant, rigid, and eventually crumbles under its own weight. Flow without structure is chaotic, unstable, and quickly dissipates into nothingness. In the modern economic landscape, when either your structural security or your financial flow is entirely controlled by external institutions, you lose the autonomous ability to build sustainable wealth for yourself. True economic liberation requires you to establish your own balance between secure asset structures and liquid financial flows.

Where This Becomes Real

These seven principles are no longer confined to the realms of ancient philosophy; they are the active programming rules of our modern reality. They are being expressed directly through our digital infrastructure, our algorithmic social networks, our centralized financial systems, and our emerging ownership models. Every single one of these networks is scaling, moving, and aligning with structural patterns that the vast majority of the population cannot see.

The Divide That Is Forming

This systemic alignment is carving a deep, invisible canyon through modern society. This divide is far more profound than a simple gap between the wealthy and the poor. It is a fundamental separation between those who possess conscious awareness and those who are trapped in a state of unconscious participation; a dividing line between the few who recognize the patterns of the world and the many who are mindlessly moved by them.

The Cost of Not Seeing

If you choose to remain blind to these foundational laws, you do not escape their influence. You will still experience their effects every single day, but you will experience them as crushing economic pressure, perpetual confusion, structural misalignment, and a constant, anxious feeling that the world is shifting in a way you do not know how to survive. Over time, this lack of structural clarity translates directly into a systematic loss of life position. It does not happen overnight; it occurs slowly, quietly, and consistently, until one day you wake up to the sober realization that you are no longer an active participant in your own life. You have been completely shaped by the system.

CLOSING

The single greatest economic risk facing the modern individual is not that the world is changing at a breakneck pace. The ultimate risk is that the world is changing according to deeply structured patterns you were never taught to see, while your daily life is being guided by them automatically. When you are guided automatically by an invisible system, the sweeping shifts of the world cease to look like unprecedented opportunities. They begin to feel like an ongoing tragedy happening directly to you.

This realization brings us to a deeper, more uncomfortable truth. Even when the system is clearly mapped out, even when the patterns are explained, and even when the opportunity is placed directly in front of them, many people will still completely miss the shift. This failure does not occur because they lack raw intelligence, nor does it happen because they lack work ethic or discipline. It happens because they were never structurally positioned to see clearly in the first place. What is deeply familiar to us is rarely examined, and the traditional path we have always followed is seldom challenged—until the day it completely stops working. This reality begs an essential question: *Why is it that so many good, hardworking people will look directly at this historic shift and never see it at all?*

Chapter 8
WHY PEOPLE WILL MISS THE SHIFT

Even When They Do Everything Right

A Story I Will Never Forget

During my years of coaching, I crossed paths with one young man whose story I will carry with me for the rest of my life. He was the absolute embodiment of everything you look for in a leader: disciplined, relentlessly focused, deeply respectful, and possessing a quiet, resilient character. He was the exact kind of player that every coach dreams of having on their roster. He was always the very first to arrive at the facility in the morning, and the absolute last to leave the field after practice. He executed every drill and handled every responsibility with flawless precision. In the classroom, he maintained a stellar record, and on the field, his performance under pressure was spectacular. He believed with his whole heart in the traditional process of success. He had mapped out a clear, honorable vision for his life: graduate with honors, secure a stable degree, and build a quiet, secure life of stability. The game of football had opened doors for him, providing him with structure, community, and opportunity, and he walked through every single one of those doors with absolute commitment and belief.

The Moment That Changed Everything

Several years after he graduated and moved on from the program, our paths crossed once again. I was excited to see him, and I asked him a simple, standard question: "*How are things going out there in the world?*" He paused for a long, heavy moment. It was not because he lacked an answer, but because the reality of his life was not matching up with the promise of the path he had been given. He had done absolutely everything right. He had checked every single box, followed every rule, and worked with relentless discipline, yet he was still struggling, still searching, and still trying to figure out why the traditional path had not led him to the security he was promised.

The Truth Most People Never Say Out Loud

Looking at him in that moment, I recognized a painful pattern that I have since seen repeated hundreds of times. This was not a failure of effort. It was not a lack of personal discipline, and it was certainly not a deficit of intelligence. It was the presence of a massive, structural gap: the disconnect between preparation and positioning. This young man had been brilliantly trained to perform at a high level inside of a structured system, but he had never been taught how to navigate the economic landscape that existed outside of it. He was trained to trade his physical energy for a grade or a paycheck, but he was never taught how to recognize the architecture of ownership, how to identify and acquire scalable assets, or how to construct something of lasting value that could continue to support his life independent of his personal labor. The educational and athletic systems had prepared him perfectly to participate, but they had left him completely unequipped to own.

Attribute	Preparation (The Traditional Loop)	Positioning (The Ownership Mindset)
Primary Activity	Executing performance within pre-existing pathways.	Observing structural design and locating leverage points.
Financial Flow	Labor → Wage → Consumption → Reset	Effort → Asset Structure → Continuous Value
Core Asset	Limited physical and mental labor-time.	Immutable ownership equity and programmatic networks.
Survival Dynamic	Reacting to economic shocks and institutional shifts.	Direct participation in systemic growth and asset yield.

And He Is Not Alone

There are millions of incredibly talented, highly educated, and deeply committed individuals living this exact story right now. They are striving every day, working long hours, and exhausting themselves, yet they are left wondering why their financial reality feels so completely disconnected from their immense effort. The hard truth of our modern economy is that effort alone does not dictate your outcome. Structure dictates your outcome. Positioning dictates your outcome. True systemic understanding dictates your outcome. Without that understanding, even the most disciplined and hardworking people can spend their entire lives running forward on an economic treadmill—exerting massive energy but never actually gaining an inch of ground.

A Different Outcome — A Second Story

But I have also witnessed a completely different kind of story, and this one is deeply personal. During my freshman year playing football for Temple University, we went on the road to play a highly anticipated game against Syracuse. They had a phenomenal, explosive wide receiver, Marvin Harrison, who wore number 8 on his jersey. I was a young safety wearing number 38. At that specific moment in time, he was just another elite opponent I had to prepare for; years later, the world would know him as an NFL Hall of Famer. Late in the first half of the game, their quarterback dropped back and tried to hit him on a deep post route across the middle of the field. I read the quarterback's eyes, broke on the ball, stepped directly in front of the receiver, and intercepted the pass.

The Moment I Almost Missed

I can still feel the electric rush of that moment—the explosion of adrenaline, the deafening roar of the crowd, and the overwhelming excitement of making a game-changing play on a massive stage. It was the exact scenario you dream about as a kid playing in the backyard. I was so completely swept up in the emotion of the play that I did not even realize what I had done with the actual football after the whistle blew. When we finally went into the locker room at halftime, I was sitting at my locker, still completely dazed and trapped in the emotional high of the interception. One of our senior safeties, a player who wore number 39 and whom I respected deeply, walked over to my locker. He had the actual game ball in his hands. He looked down at me, handed me the ball, and said: "*Here you go, Duff... you keep that.*" It was a simple, quiet gesture, but it carried a profound lesson.

What He Saw That I Didn't

In that moment, the interception was just a fleeting, exciting play to me—a temporary emotional high. But my teammate saw something entirely different. He possessed the rare ability to recognize long-term value in a moment that I was only experiencing emotionally. He understood that this physical ball was not just a piece of sports equipment; it was a permanent asset, a physical anchor of a historic moment that would hold value for the rest of my life. Years later, I watched that teammate's life unfold after his athletic career came to a close. He transitioned from sports into the business world, then into commercial real estate development, ultimately becoming a respected owner and developer in the hospitality industry. Through the company he founded, he has helped develop and acquire projects across multiple states, with a reported development pipeline valued at more than $1.1 billion. His success did not happen by random chance or luck; it was the natural result of consistent systemic alignment.

The Difference Between the Two Stories

Both of these individuals possessed the same elite work ethic, the same fierce discipline, and the exact same level of personal commitment. The divergence in their lives was not determined by their effort, but by their relationship to awareness. It was dictated by their ability to recognize true value in real time, and their understanding of how to transition from participation to structural ownership. That is the ultimate difference. It is never about how hard you work; it is about what you choose to align your work with.

Why Good People Still Miss the Shift

This realization brings us to a highly uncomfortable economic truth. Good, honest people do not miss the great shifts of history because they are lazy or undisciplined. They do not miss them because they lack intelligence or desire. They miss them because their minds have been systematically trained to see the world through an outdated lens. They were taught by well-meaning parents and institutions to work hard, stay consistent, and follow the traditional path. While those traits build beautiful human character, they are completely silent on how to read, navigate, and own the very systems you are working inside of.

The Trap of Doing Everything Right

This is the ultimate systemic trap that no traditional school or institution prepares you to face. You can perform every duty perfectly, follow every rule to the letter, and still end up economically devastated and misaligned. This tragedy occurs because the very definition of what is "right" was engineered inside of an economic system that is now rapidly disappearing. If the foundational architecture of the world undergoes a massive transformation, but your personal blueprint for success remains completely unchanged, you will find yourself executing a flawed plan with perfect discipline—marching headfirst in the wrong direction.

The Pattern Most People Never Recognize

Major historical transformations never announce their arrival with a grand public celebration. They begin in the quiet, dusty corners of innovation. New ideas emerge, experimental systems are tested, and entirely new infrastructures are quietly constructed beneath the surface of daily life. Meanwhile, ordinary life continues as it always has: people go to work, pay their bills, and shoulder their daily responsibilities. Nothing feels radically different on the surface—until the day the transformation reaches critical mass, and suddenly everything is different. By the time the average person realizes the world has changed, the new economic structures are already locked into place, and the windows of early positioning have closed.

Information Is Not Awareness

We are currently living in an era of unprecedented information overload. We are continuously bombarded by non-stop news updates, social media content, and endless digital noise. But you must never mistake the accumulation of information for the possession of true awareness. People hear about major technological shifts and financial changes on the news every day, but they do not truly see them. They perceive isolated fragments of change without ever recognizing the massive, connected pattern that is forming directly beneath them. That single cognitive gap is the difference between economic survival and systemic obsolescence. True awareness is the only force capable of producing strategic positioning.

The Hard Truth About Work

Right now, we are witnessing highly educated, deeply experienced professionals with advanced degrees and elite credentials who are completely unable to find economic stability. They are drowning in uncertainty, despite having done everything they were told to do. This crisis is occurring because we are not living through a temporary economic downturn; we are living through a permanent structural shift. The role of human labor is being systematically reduced across every industry on earth, driven not by human hands, but by autonomous systems.

Artificial intelligence, advanced automation, and robotics are scaling at an exponential rate. As these technologies continue to evolve, the economic value of traditional labor pathways is rapidly evaporating. A formal degree or a specialized skill set still holds value, but they are no longer sufficient to guarantee your survival on their own. The global economy is shifting its center of gravity away from human labor and directly toward the ownership of assets.

The Question No One Is Ready For

There is an existential economic question looming on our immediate horizon, and when it fully arrives, it will not be an academic debate. It will be a visceral, daily reality. *What happens to human society if traditional work completely disappears*?

Imagine a world where automated, digital systems produce every physical good, manage every supply chain, and execute every service—faster, cheaper, and with flawless precision, without requiring human labor. At first glance, this scenario sounds like the ultimate liberation of humanity. But if you look deeper at the architecture of our current system, the illusion shatters. Our entire society is built upon a single, rigid linear loop:

Work → Income → Survival

If you completely remove work from this equation, the system does not magically adjust itself; it completely breaks.

The Systemic Collapse Model

When we analyze the transition from traditional economic framework systems to the incoming age of total automation, the breakdown of our societal loops becomes starkly apparent.

Economic Framework	Core Operational Input	Intermediary Link	Terminal Outcomes
Traditional Loop	Human Labor	Financial Wages	Individual & Societal Survival
Automated Reality	Autonomous AI & Robotics	**[Severed Wage Link]**	Severe Systemic Economic Crisis

What Comes After That

When that traditional structure breaks, the ultimate question of human civilization changes forever. It is no longer a question of who has the best job or who works the hardest. The question becomes:

- *Who owns the automated systems of production?*
- *Who controls access to the digital networks?*
- *Who participates in the immense value being generated?*

If the volume of global value continues to grow exponentially through technology, but human labor is no longer required to produce it, the gap between those who own the assets and those who trade their time will not remain static. It will widen into an insurmountable, permanent canyon.

This is the sobering, ultimate truth of our era. Millions of good, honest, hardworking people will completely miss this monumental shift—not because they failed in their efforts, but because they were never shown how to see the design of the game. They will continue to work harder, strive longer, and do absolutely everything "right" inside of an economic system that no longer values or rewards what they were trained to give. By the time they look up from their labor, the new digital structures will be fully established, the prime positions will be permanently occupied, and the wealth of the future will already be moving through networks they do not own.

This reality means that simple intellectual awareness is no longer enough to save us. We must choose a definitive direction. If our childhoods trained us for passive participation, then our adulthood must be dedicated to active ownership. If ownership must be learned to survive, then we must construct a clear, accessible pathway to acquire it. We need a practical, step-by-step methodology to transition our lives from passive participation to strategic positioning, and ultimately, to true structural ownership. This journey cannot be left to blind luck or guesswork; it must be mapped out with absolute structural precision.

And that practical path is exactly what we are going to build next.

Chapter 9
THE OWNERSHIP PATHWAY™

From Participation to Ownership

The Pattern Behind Every Outcome

Every system in our modern economy is continuously producing value. Day in and day out, the machinery of commerce hums, transactions are processed, platforms are utilized, and networks expand. Yet, despite this constant generation of wealth, we must observe a critical and undeniable truth: not everyone participates in that value in the same way. The economic positions human beings occupy within these systems are vastly different. Some individuals contribute their time and energy merely to keep the wheels turning. Others benefit from the surface-level convenience and products these systems yield. A rare few, however, actually own the underlying infrastructure itself.

The profound difference between these systemic positions is not a matter of random chance, luck, or simple inheritance. It follows a predictable, structured pattern—a specific sequence of developmental phases that we call **The Ownership Pathway™**. This framework maps the journey from passive compliance to systemic leverage. It is the roadmap of economic self-determination that most people are never shown by traditional educational institutions or societal structures.

Pathway Stage	Core Operational focus	Cognitive Shift	Ultimate Economic Position
1. The Ownership Gap	Recognizing systemic disparity and the limits of labor-only contribution.	**From:** Unconscious compliance **To:** Critical observation	Realizing that trading time for money is a structural loop with a hard ceiling.
2. Awareness	Observing macro shifts, technological transitions, and emerging ledgers.	**From:** Passive consumption **To:** Active education	Recognizing that the rules of the economic game are undergoing a fundamental reset.
3. Positioning	Locating early opportunities and aligning personal skills with structural changes.	**From:** Reactive panic **To:** Proactive preparation	Strategic placement near emerging value hubs before they become obvious.
4. Participation	Engaging directly with new networks, digital assets, and local initiatives.	**From:** Sideline observation **To:** Strategic action	Building hands-on competence and deploying capital or labor inside the new loop.
5. Ownership	Securing equity, holding programmable assets, and capturing long-term value.	**From:** Transactional earning **To:** Structural compounding	Achieving systemic autonomy where assets generate value independent of physical labor.

1. The Ownership Gap — Seeing What Was Missing

The first necessary step on this journey is confronting the reality of the Ownership Gap. For generations, the vast majority of the population has been brilliantly trained to perform tasks, execute strategies, and function seamlessly inside systems—yet they have never been taught how to own them. This disconnect is not a failure of individual effort, intelligence, or personal character. It is a fundamental gap in exposure and systemic education.

It is the unseen distance that separates mere participation from true ownership, traditional income from compounding equity, and temporary institutional access from permanent systemic control. Until this gap is clearly identified and personally acknowledged, an individual will remain trapped in the labor loop, working harder and longer while the true value they generate continuously flows upward to those who own the infrastructure.

2. Awareness — The Moment Everything Shifts

Awareness is the precise spark where the journey of self-determination begins. It is the sudden, clear-headed realization that the economic systems surrounding us are not static monuments, but living networks undergoing a profound transformation. In this phase, you begin to see that the historical rules of property, currency, and value are being systematically rewritten.

Without this initial cognitive awakening, there can be no meaningful movement toward financial sovereignty. However, we must also acknowledge that awareness alone is insufficient. Simply observing the flaws of a system or recognizing the arrival of a new technology does not automatically alter your position within it. You cannot think your way across the gap; you must use your awareness as a guide for strategic action.

3. Positioning — Deciding Where You Stand

Once you have achieved awareness, the crucial question shifts from *what is happening* to *where do I stand in relation to what is unfolding*? Positioning is not about possessing massive amounts of starting capital or having instant access to elite rooms. Rather, it is the disciplined process of preparing yourself intentionally in the quiet, early hours of a paradigm shift.

It involves actively seeking out systemic literacy, building deep relationships with forward-thinking builders, understanding the macroeconomic cycles of timing, and aligning your unique skill sets with emerging opportunities. Positioning is quiet, unglamorous, and almost entirely invisible to the distracted masses. Yet, it is the single most critical factor that determines your ultimate trajectory when the transformation finally reaches critical mass.

4. Participation — Moving Into the System

Participation is the decisive transition from intellectual contemplation to direct physical action. It is the moment where theory becomes practice, and knowledge is converted into tangible motion. In this phase, individuals and local communities actively engage with new networks, explore emerging asset classes, collaborate with other early adopters, and contribute real value to the systems being built.

Participation is where you gain real competence, test your understanding under real-world conditions, and build momentum. However, we must remain highly cautious: participation alone does not guarantee long-term ownership. Throughout history, millions of people have participated in massive economic booms—acting as the early users, creators, and laborers—without ever securing a permanent stake in the systems they helped popularize.

5. Ownership — Alignment With Value

Ownership is the ultimate destination of the pathway. It is far more than the simple physical possession of material goods; it is a profound structural alignment with the engines of value creation. True ownership means possessing an undeniable stake in the infrastructure, holding a defining voice in its future governance, and maintaining a permanent, compounding connection to the long-term outcomes of the network.

This is the precise space where financial stability is anchored, where wealth compounds organically, and where generational legacies are securely constructed. But ownership is also a deeply internal reality. Two individuals can stand in the exact same historical moment, observe the identical structural opportunity, and walk away with completely opposite life outcomes. This divergence occurs because one possessed the internal identity of an owner, enabling them to see and capture the structural value that the passive participant simply overlooked.

The Hidden Lever — Perspective

Ultimately, your perspective dictates your style of participation, and your participation dictates the level of ownership you can sustain. Perspective is the hidden cognitive lever that determines how far and how fast an individual can move along this developmental pathway.

The Ownership Pathway™ Flow

The Gap → Awareness → Positioning → Participation → Ownership

Turning the Framework Into Action

Most people allow their journey to stall at the level of passive understanding. They read the books, watch the shifts, and discuss the theories, but they never bridge the gap into execution. To prevent this stagnation, we must actively translate the conceptual framework of the Ownership Pathway™ into an immediate, repeatable action system.

Step 1 — Identify the System

- **The Diagnostic:** Pause and look closely at the primary economic network you are currently operating inside of.
- **The Questions:** *Who ultimately owns the infrastructure of this system? How is value generated here, and who captures the surplus at the end of the day?*

Step 2 — Build Awareness

- **The Diagnostic:** Lift your gaze to the horizon to study the macro shifts occurring around your industry.
- **The Questions:** *What structural changes are currently unfolding in this system? What new technological or regulatory frameworks are taking shape, and what is becoming more valuable as a result?*

Step 3 — Position Intentionally

- **The Diagnostic:** Locate the active hubs of early innovation and systemic transition.
- **The Questions:** *Who already understands this space at a high level? What are they doing differently, and how can you align your attention, education, and connections with them today?*

Step 4 — Participate Strategically

- **The Diagnostic:** Identify low-risk, high-learning entry points where you can gain active experience.
- **The Questions:** *Where can you actively engage with this emerging system today? What unique value can you contribute, and what is your immediate, practical entry point?*

Step 5 — Align With Ownership

- **The Diagnostic:** Transition your focus from transactional income to compounding equity.
- **The Questions:** *How can you convert your daily efforts into long-term equity or programmatic assets? What are you actively building today that will continue to grow and yield value independent of your physical time and labor?*

Closing

The path to economic sovereignty has never been hidden; it was simply never explained to us by the institutions designed to produce compliant labor. The traditional world was engineered to teach you how to participate as a consumer and a worker, while remaining completely silent on how to move beyond those positions. But once you see the architecture of the Ownership Pathway™, it becomes impossible to unsee it. You begin to recognize its presence in every market cycle, every emerging technological platform, every community initiative, and most importantly, within yourself.

The global system is already changing at an exponential rate, and the new digital infrastructure of the world is being laid down as we speak. The difference between those who benefit from this historic transition and those who are left behind will not be determined by effort alone. It will be determined by position.

THE OWNERSHIP PATHWAY™
From Participation → Ownership
THE OWNERSHIP GAP
• Unseen Barrier •
• Participation ≠ Ownership
• Income ≠ Equity
• Access ≠ Control
AWARENESS
'Seeing the System Clearly'
• Systems Changing
• Opportunities Emerging
• Rules Flexible
POSITIONING
'Preparing for Alignment'
• Learning
• Aligning
• Building Relationships
• Understanding Timing
PARTICIPATION
'Engaging with the System'
• Taking Action
• Collaborating
• Exploring Opportunities
• Contributing Value
OWNERSHIP
'Alignment with Value'
• Stake
• Voice
• Long-Term Outcomes
• Stability & Legacy
PERSPECTIVE & OUTCOMES
LIMITED PERSPECTIVE
Passive Participation
Short-Term Outcomes
EXPANDED AWARENESS
Strategic Participation
Long-Term Ownership
Perspective shapes participation.
Participation shapes ownership.
EMPIRE NEXUS
MAKING THE UNSEEN - UNDERSTOOD

Chapter 10
DIFFERENT SIDES OF THE SAME COIN

Perspective, Pressure, and the Beginning of Awareness

The Power of Identity

During this past (2025-2026) college football season, the sports world was treated to a vivid, electrifying glimpse of what Miami Hurricanes football felt like during its legendary reign in the early 1990s. Whenever that iconic Hall of Fame wide receiver—the legendary number 88 of the Dallas Cowboys, who now commands the microphone as a prominent broadcaster—speaks about his alma mater, he brings that historic era back to life with a fierce, burning passion. He speaks of the relentless energy, the unshakable confidence, and the fierce, unapologetic belief that defined their locker room.

I have always held an immense amount of respect for that energy, because it was never just empty locker-room talk—it was their core identity. If you were a part of college football back in those days, you did not need that culture explained to you. You didn't read about it in the sports pages; you felt it deep in your chest the moment you stepped onto the field.

During my years competing in the Big East Conference in the early to mid-1990s, the Miami Hurricanes were a completely different breed of competitor. They possessed an unmistakable identity, a unique style of play, and a physical presence that was entirely their own. This swagger was not a performative act they turned on for game day; it was woven into the very fabric of who they were as human beings. And they were spectacular.

Year after year, that program acted as a dominant factory for NFL talent, producing legendary Hall of Fame players, and even a future superstar wrestler who would go on to become the biggest movie star in the world. They talked differently, they walked differently, and they carried themselves with a towering confidence that the outside world often mischaracterized as pure arrogance. But beneath the swagger lay a simple, undeniable truth: they could back up every single word they said.

Now, do not get me wrong—we had elite players on our roster at Temple University, too. We were tough, disciplined, and hungry. But Miami... Miami was *them.*

Standing in the Orange Bowl

I will never forget that hot Saturday afternoon at the historic Orange Bowl Stadium in Miami, Florida. As a kid growing up playing football in the backyard, the Orange Bowl was a sacred, mythical place I had dreamed about for years. And now, standing on that legendary grass field, the reality washed over me: I was finally here, living inside the dream.

It was a 1:00 PM kick-off, and the temperature on the field had already soared past a blistering 100 degrees. It was the kind of thick, heavy South Florida heat that sits like a lead weight on your shoulders before the opening whistle even blows. You could feel a intense, crackling tension in the air—a pressure generated not just by the brutal weather, but by the massive weight of the moment itself.

All week long, the media noise had been building to a deafening crescendo. The national headlines, the local conversations, and the sports commentary had framed the matchup as a clash of civilizations. This was not just another game on our schedule; we were lining up against the self-proclaimed "bad boys" of college football, and the world was watching to see how we would respond.

Perspective Map (1990s): The Orange Bowl Field

Attribute	Temple University (The Hungry Safety)	University of Miami (The Relentless #52)
Role	Active Defense	Observed Greatness
State	Brutal Labor	Systemic Flow

Going into that grueling game, our secondary at Temple was highly respected, ranked among the very best passing defenses in the entire country. But fate dealt us a difficult hand that week. Two of our key starting defensive backs—both of whom, ironically, were proud Miami natives—were out with injuries. The immense responsibility of defending the deep field fell squarely on our shoulders. As the starting free safety, I was the designated quarterback of our defense. I was the leader who had to see the entire field, make the adjustments, and hold our defensive line together in the face of an absolute storm.

The Witness of Greatness

I do not remember every single play or every down from that afternoon, but I remember with absolute, crystal clarity how I felt. On that day, I was not merely competing inside the game; I was actively witnessing something historic.

Years later, as a collegiate coach, I would help guide and develop elite young athletes who reached the absolute pinnacle of the NFL. But even before I began coaching, I was constantly surrounded by that level of talent. All three of the linebackers I played alongside at Temple were drafted directly into the NFL. Over the course of my football journey, four players deeply connected to my defensive path went on to play professionally. Two of them earned coveted Super Bowl rings—one as an elite defender on the field with the legendary Tampa Bay Buccaneers defense, and another as a highly successful executive in the front offices of the Green Bay Packers, and the Kansas City Chiefs. These were high-character, fiercely disciplined men who thoroughly deserved every ounce of success they achieved.

But that hot afternoon in the Orange Bowl always pulls me back to the grass. I remember standing deep at free safety, completely locked into my responsibilities. Yet, whenever our Temple offense took the field, my perspective shifted. Instead of resting on the bench, I found myself standing on the sideline, completely mesmerized by Miami's defense.

Specifically, my focus locked onto one linebacker. He was incredibly fast, impossibly strong, and utterly relentless in his pursuit of the ball. He wore number 52. At that precise moment in time, he was just another talented player on a roster stacked with future pros. Years later, the entire world would recognize him as one of the most ferocious and transformative defensive players in the history of the National Football League: Ray Lewis.

There were moments during that game where I should have been resting or reviewing our defensive coverages, but my eyes remained glued to number 52. It was not a distraction; it was a profound, quiet recognition. Greatness has an undeniable, magnetic quality that commands your attention, forcing you to stop and observe, regardless of which sideline you stand on.

Later in my life, I would experience that exact same feeling of recognition from the opposite side of the whistle—watching young men I coached step fully into their own greatness. The role was entirely different, but the core awareness was identical. That was the moment I understood a deeper truth about the human experience: sometimes you are actively in the arena, competing; other times you are outside the arena, witnessing. Both perspectives are valuable, and both are necessary pieces of the exact same puzzle.

The Question of the Coin

That brutal game gave me far more than intense competition and physical exhaustion; it gave me a permanent shift in awareness. But if I am being completely honest, I wanted to beat Miami so desperately that the contest transcended the boundaries of a sport. It felt deeply, intensely personal.

To fuel my competitive fire, I told myself standard athletic lies: *I hate how they walk. I hate how they talk. I hate how they carry themselves on the field.* But as the years passed and the heat of youth faded, I realized it was never hatred. It was pure, unadulterated recognition of a standard I aspired to reach.

At the end of that exhausting afternoon—after we had poured every single ounce of our physical souls onto that field—our head coach walked over, picked me up, and looked me in the eyes. "*You gave everything you had today, Duff,*" he said. And he was right. We had given everything, and we still lost the football game.

That afternoon was the very first time in my athletic career that my entire body completely locked up in agonizing, full-body muscle cramps. There was absolutely nothing left in the tank. I had to receive emergency IV fluids in the locker room immediately after the game, more IVs on the charter plane back to Philadelphia, and spend hours undergoing recovery treatment in the training room. I had pushed my physical body to the absolute brink of collapse, all for a single chance to win a game against an elite system.

Years later, I had to ask myself a very real, uncomfortable question: why did I feel such a intense, complicated emotional charge toward that team?

There was a talented kid from Buffalo on that loaded Miami squad. I had always respected his journey, and years later, our paths would cross again as our own sons grew up playing in some of the same youth athletic leagues in Western New York. Whenever we met, we never talked much about those intense college afternoons in the 1990s—but we did not have to. There was a quiet, unspoken understanding between us. We both knew what we had survived.

Back in 1990's, the players on that legendary Miami roster were so deep that he did not see much playing time; he worked primarily on special teams, while I was a three-year starter and the captain of our secondary defense. Over the years, a curious thought crossed my mind: *If that kid could have switched places with me back then—to be the starting safety, to play every defensive snap, to command the field—would he have taken that trade?*

Then, I turned that same difficult question back on myself: *If I could have switched places with him—to be a part of that legendary Miami program, to absorb that unstoppable championship energy, to share that unbreakable brotherhood, even if it meant playing far fewer snaps—would I have taken it?*

That was the moment the ultimate truth of the experience hit me. This game was never about who was individually better. It was not even about the final score on the scoreboard. It was about perspective. In the grand design of that historic afternoon, we were both there, standing on the exact same world-class stage, experiencing the same event from two entirely different viewpoints. We were different sides of the exact same coin.

Looking Beyond the Division

To this day, whenever I see him around the city of Buffalo, there is a profound, quiet respect between us. No elaborate words are ever required, because we both lived the reality of the pressure. And that lesson stayed with me long after I played my final down of football.

When I look at the modern world today, I see the exact same pattern of division, tribalism, and labeling repeating itself across every layer of our society. We have constructed a culture entirely centered on artificial sides: Democrat versus Republican, Liberal versus Conservative, Us versus Them, Right versus Wrong. We have adopted a highly aggressive, binary language that forces people to retreat into their ideological corners, treating anyone on the opposite side of the field as an existential enemy.

But that divisive language does not move our civilization a single inch forward. The truth of the human experience is far simpler than the media narratives suggest: we are all living inside the exact same historical moment, navigating the same systemic pressures, just from different structural perspectives.

Just like that blistering afternoon in the Orange Bowl, we have different roles, different starting positions, and different responsibilities. But we are still operating inside the exact same global economic system.

One day, when the dust of this modern transition finally settles, we will all look back at this era of rapid change with the same clarity of hindsight. We will remember how we chose to show up, how we trained our minds to think, how we treated those on the opposite side of the field, and how we ultimately played the game.

What we are living through right now is not a temporary economic downturn or a passing cultural trend. It is a permanent, exponential acceleration of history. Our global systems are shifting, our financial structures are evolving, and the pace of life is increasing at a speed that human biology was never designed to handle. We are all inside the pressure cooker together.

So, before you blindly choose a side, before you join an ideological tribe, and before you react in anger, pause. Look inward and ask yourself a simple, piercing question: *Are you standing in what you believe out of calm, analytical clarity—or are you simply reacting to the fear and outrage being algorithmically fed to you?*

Your answer to that question will shape every single decision, position, and outcome that follows. When that level of self-awareness finally settles into your consciousness, a profound shift occurs. You stop viewing people through the shallow lens of division, and you begin to see the underlying architecture of the world for what it truly is.

You begin to recognize that every global system, every local environment, and every meaningful opportunity is built upon a single, foundational reality: **Assets**.

Not paper transactions, not volatile market speculation, and not superficial political debates—but the real, underlying assets that generate value and make human life possible. Once you achieve this level of clarity, it becomes impossible to look at the world the same way again. You begin to ask the only question that matters for your survival: *What are the actual assets shaping the world around me, and who owns them?*

Chapter 11
REAL-WORLD ASSETS

Seeing What Was Always There

The Surface Illusion

Look around you. Do not do it quickly, and do not do it in the distracted, hurried manner that defines how most people move through their daily lives—rushing from one task to the next, staring at screens, without ever truly perceiving the physical environment surrounding them. Pause for a moment. Take a deep, intentional breath, and actively observe the world you inhabit.

Notice the local restaurants serving multi-generational families. Study the brick-and-mortar buildings lining the streets of your neighborhood. Listen to the music moving through the speakers in your car or the headphones in your ears. Watch the small business owners unlocking their doors before sunrise, and observe the community programs working tirelessly to help young people find structure and direction in life.

At first glance, this looks like mere human activity. It appears to be nothing more than random movement, physical energy, and the standard noise of daily survival. But that is only the superficial layer of reality. When you slow down, step back, and train your mind to observe instead of react, a much deeper, more organized structure reveals itself. This is not a chaotic collection of independent events; it is a highly coordinated system of value in motion.

The Multi-Dimensional Asset

The Visible Surface — Everyday transactions, jobs, and physical buildings that people interact with daily.

The Unseen Structure — The underlying asset: cash flows, yield, equity, and intellectual property rights.

Every single physical space you pass, every digital network you interact with, and every local system you move through is constructed upon a foundation that the vast majority of the population never stops to consider: **Assets**.

Assets are the quiet, foundational structural pillars of human civilization. They are often completely invisible to the untrained eye, operating silently in the background of society. An asset is not merely a physical object or a balance sheet entry; it is a legally and technologically structured engine that holds real-world value, produces ongoing opportunity, organizes local economies, and dictates the long-term trajectory of entire communities without ever needing to announce its existence. Once you learn how to identify these engines, your relationship with the world changes permanently. You realize that the economy is not just happening to us; it has been systematically built.

The Engines of Community Value

Every city, town, and neighborhood is a layered, highly complex ecosystem of assets. These engines take many forms: private businesses, commercial real estate, global brands, intellectual property, physical infrastructure, and creative masterworks. Even structured community programs that consistently transform lives and elevate local families carry immense, measurable value that extends far beyond a single moment in time.

These are not random, disconnected pieces of property. They are the primary engines of societal health. They create stable jobs, generate ongoing municipal revenue, influence cultural trends, and anchor the stability of entire neighborhoods. Over time, these assets accumulate leverage, establishing the foundational continuity that allows families to build sustainable lives across generations.

Yet, most people move through these systems without ever recognizing their structural design. To the passive participant, a local business is viewed as "just a place to work." A global brand is dismissed as "just a cool logo." A piece of real estate is categorized as "just a building."

But when your cognitive perception shifts, everything changes. What once looked ordinary reveals its true nature as a powerful structural engine—a reservoir of value that has the capacity to grow, scale, and shape human outcomes far beyond the present moment. This is where your awareness must deepen. Before you can ever understand the mechanics of ownership, you must first train your eyes to see where value actually lives.

The Foundation Beneath Everything

Every economic system, every local government, and every individual opportunity rests upon the bedrock of assets—not opinions, not passing trends, and not political narratives. Real-world assets are the ultimate truth of any economy.

- **A viable business** generates consistent operational revenue.
- **A commercial building** produces reliable rental income.
- **Intellectual property** yields ongoing royalty flows.
- **Physical infrastructure** enables industrial scale and connectivity.

Even systems that do not appear financial on the surface—such as schools, grassroots nonprofits, and community mentorship programs—are actively creating immense, measurable value. They shape human capital, influence long-term societal outcomes, and build the future leaders of our communities.

The critical divide in our society is simple: some value is legally recognized, mathematically measured, and securely owned, while other immense value is created every single day by ordinary people but remains unseen, unstructured, and completely disconnected from ownership.

This structural distinction is everything. When value is made visible, it can be measured, protected, supported, scaled, and owned. When it remains invisible, it is still created by human labor—but it is captured entirely by those who control the master ledgers of the system.

So, the fundamental question of our era begins to shift: *Who owns these underlying assets, and who truly understands how they work?*

True ownership is not just a matter of having legal access; it is a matter of active cognitive awareness. Two people can walk past the exact same commercial building in downtown Buffalo. One sees nothing more than a structure of brick and mortar. The other sees a highly lucrative real-world asset—generating consistent yield, holding generational value, and positioned strategically for future growth.

Two people can engage with the exact same digital platform. One participates as a passive consumer, trading their attention for temporary entertainment. The other analyzes the structural design of the platform, positioning themselves to capture a portion of the value flow.

This single cognitive difference is where human outcomes begin to separate. Once assets become visible to you, the world stops looking like a collection of random events. It reveals itself as a structured, intentional, and highly connected design. You begin to realize that assets do not have to exist solely in a heavy, physical format; they can be represented, structured, and connected across modern digital networks in ways that were completely impossible just a decade ago.

This realization leads us directly to the next phase of our journey. If real-world assets are the absolute foundation of economic value, then the urgent question is no longer just what they are—but how their ownership structures are actively evolving.

Income vs. Assets

There is a quiet, systemic misunderstanding that shapes how the vast majority of well-meaning people move through their lives. They spend their entire careers focusing exclusively on income. Income is essential for immediate survival; it pays the bills, purchases groceries, and allows you to participate temporarily in the consumer economy.

But income alone has no power to shape your economic future. Assets do.

The Value Divergence Model

Path	Flow
Traditional Focus	Labor → Income → Survival (Reset)
Ownership Focus	Effort → Assets → Legacy (Growth)

A business owner is not merely working to earn a salary; they are actively building an engine of equity that can grow, expand, and outlast their physical capability. A property owner is not just collecting a monthly check; they are securing a tangible source of long-term leverage and wealth. This is the ultimate distinction: income supports your life today, but assets secure your legacy tomorrow. Once you internalize this truth, it becomes impossible to navigate the economy the same way again.

A Simple Example — A Deeper Truth

To understand this concept in its simplest form, imagine an old, vacant building sitting quietly on a corner in your city. For years, the distracted crowds walk past it, seeing nothing more than an eyesore, a useless monument to a bygone era.

Then, one day, an individual with an ownership mindset stops and looks at the building. They do not see what it is; they see what it can become. They acquire the property and begin to restore it.

Slowly, inside that once-dead space, a new community ecosystem begins to take form:

- A local coffee shop opens on the ground floor, creating jobs and neighborhood connection.
- An independent business moves in next door, generating local economic activity.
- A grassroots nonprofit establishes educational programs on the second floor, mentoring youth.
- Local families move into the renovated apartments above, finding a stable place to live.

What actually changed? It was not just the physical brick and mortar of the building. It was the structural meaning and utility of the space.

The building was transformed from a dormant physical shell into an active, productive real-world asset. It became a vital point of human connection, a source of stable opportunity, and an active center of local economic movement. That single structure now exerts a positive, cascading influence on the lives of everyone around it—shaping jobs, supporting families, and guiding the direction of the community. This is what assets do. They do not just exist; they actively shape the human experience.

A Moment of Pause

At this point in our conversation, I want you to temporarily step away from this text. Open a browser window, step outside of your immediate environment, and perform a series of simple searches:

- *Search:* **BlackRock real-world asset tokenization**
- *Search:* **J.P. Morgan Onyx blockchain network**
- *Search:* **Ripple enterprise tokenization solutions**

The goal is not to memorize the technical jargon or become an overnight computer science expert. The goal is to recognize a simple, monumental fact: the largest financial institutions and custody systems on earth are actively moving trillions of dollars of real-world value onto programmable digital ledgers.

This is not a theoretical experiment being conducted in academic circles. The very systems that organize, verify, and enforce property rights for global civilization are undergoing a profound evolution. New technological frameworks are emerging—not to artificially create new value, but to organize, represent, and move existing real-world value with unprecedented efficiency.

At the absolute center of this global paradigm shift is a concept that will completely define the next era of human history: **Tokenization**.

What Tokenization Really Means

The underlying premise of tokenization is incredibly elegant: it is the process of representing authentic ownership of a real-world asset through secure, decentralized digital systems. In many cases, this technology allows high-value assets to be divided into highly precise fractional units, enabling broad participation from everyday people who were historically locked out of these markets.

This technology introduces an unprecedented level of systemic access. But we must remain deeply grounded: access alone is not the ultimate outcome. Technology does not possess the magical power to create value out of thin air; it simply organizes and distributes it.

Once you understand this distinction, the conversation around digital assets changes completely. The future is not about volatile speculation or digital hype; it is about clearly seeing the physical assets of the world, understanding how they are being digitized, and positioning yourself to participate in their ownership before the new infrastructure is locked into place.

Chapter 12
TOKENIZATION

A New Way to Represent Ownership

The Foundation of Value

A profound, quiet transformation is currently unfolding across the global economic landscape. It is not happening loudly in the sensationalized headlines of mainstream media, nor is it always visible to the average person navigating their daily routine. But beneath the surface of global commerce, the very mechanics of how human beings organize, verify, and transfer value are being fundamentally rebuilt.

To truly comprehend this shift, we do not need to begin with complex computer code, cryptographic algorithms, or technological jargon. We must start with a much simpler, more ancient concept: **Ownership**.

Before any asset can be tokenized, it must first exist in reality. A commercial building, a growing private business, a fertile plot of land, a globally recognized brand, or a revolutionary scientific idea—all of these are real-world assets. They are not new innovations; they have existed for centuries as the absolute bedrock of human wealth.

What is currently changing is not the physical existence of these real-world assets, but the architectural medium we use to represent and prove who owns them.

The Medium of Representation

Era	Medium
Ancient	Clay Tablets / Physical Ledgers
Industrial	Paper Deeds / Stock Certificates
Digital	Centralized Databases / SQL Servers
Programmable	Distributed Ledgers / Tokenized Assets

From Paper to Digital

For the vast majority of modern economic history, human property rights have been recorded and enforced using mediums that are highly familiar to us: paper documents, physical contracts, hand-signed deeds, and centralized institutional ledgers.

- **A paper deed** proves your legal ownership of a home.
- **A stock certificate** represents your fractional ownership of a corporation.
- **A written contract** defines the boundaries of your rights and responsibilities.

These paper-based legacy systems worked reasonably well, providing the foundational stability that allowed the industrial and financial eras to scale globally. But as our society becomes hyper-connected, these traditional systems reveal massive, structural limitations.

They are incredibly slow, requiring days or weeks of manual verification by expensive intermediaries. They are highly restricted, often requiring immense starting capital and elite connections to participate. They operate in isolated, siloed databases that cannot communicate with one another, creating massive friction and inefficiencies across global markets.

We now live in a world where ideas scale globally in seconds, where markets operate continuously across borders, and where communities are no longer defined by physical geography. When these archaic, paper-based systems meet a world moving at the absolute speed of light, the friction becomes unbearable. The system must evolve.

What Tokenization Actually Is

Stripped of all the speculative noise and marketing hype, tokenization is a remarkably straightforward structural process. It is the technological method of representing authentic ownership of a real-world asset as a secure, programmable digital token on a distributed ledger.

That is the entire definition. It is not a speculative investment scheme; it is a highly advanced form of legal and financial infrastructure.

Instead of your ownership of an asset existing solely on a paper document locked in a county registry or in an isolated bank database, that ownership is recorded on a transparent, tamper-proof digital ledger. It can be instantly tracked, continuously verified, and seamlessly transferred globally without the need for slow, centralized middle-men.

The Tokenization Value Loop

Real Asset (e.g., Land) → Tokenization → Digital Token (Programmable) → Redemption → back to Real Asset

Furthermore, because these digital tokens are infinitely divisible, they allow ownership to be broken down into highly precise fractional shares. This introduces a revolutionary structural possibility: the complete democratization of high-value asset ownership.

- **A commercial office building** that once required fifty million dollars of capital can now be structured to allow fractional participation from thousands of everyday investors.
- **A vital piece of municipal infrastructure** can be connected directly to the local community members who use it every day, aligning public utility with direct economic benefit.
- **A creative masterwork** can be structured to track intellectual property rights globally, automatically distributing micro-royalties to the creators in real-time.

The physical nature of the underlying asset does not change. The business still has to produce, the building still has to stand, and the infrastructure still has to function. What changes is the efficiency, transparency, and accessibility of its ownership structure.

Why This Matters

This paradigm shift is not a simple debate about technological tools; it is a monumental shift in societal access. The manner in which property rights are structured has always dictated who is allowed to participate in the wealth of a nation.

For generations, the most lucrative, stable, and compounding asset classes on earth—such as commercial real estate, private equity, and sovereign debt—have been locked behind high institutional walls. They required massive amounts of capital, specific regulatory clearances, and elite institutional connections. Consequently, the vast majority of the population was structurally barred from participating in the growth of these assets, leaving them with no choice but to trade their finite labor for a depreciating fiat currency.

Tokenization systematically dismantles these historical barriers. It creates a highly efficient, transparent digital layer between real-world value and everyday human beings. This technology makes ownership visible, transparent, and highly flexible, allowing the average person to transition from a passive consumer to an active, structural owner of the world around them.

Once this level of structural clarity is achieved, your relationship with the economy changes. You stop asking, "*What do I have access to?*" and you begin asking, "*How is access to this asset being structured, and where does the value flow?*"

What Does Not Change

To successfully navigate this economic transition, we must remain fiercely grounded in reality. This is the precise point where many early participants lose their way, swept up in the speculative hype of digital trends.

Tokenization does not possess the magical power to create real-world value out of thin air. It is an infrastructure, not a creator.

- **A failing business** will still fail if it is tokenized.
- **A poorly managed building** will still lose value if it is digitized.
- **A useless idea** remains useless, regardless of how many digital tokens represent it.

Technology does not replace the hard work of building and maintaining value. The operational business must still be run with excellence, the real estate must still be managed properly, the creative work must still resonate with people, and the community must still be actively cultivated.

The digital token is simply a mirror. If the underlying asset is valuable, the token represents that value with perfect efficiency. But if the underlying asset is garbage, the token simply allows you to trade that garbage at the speed of light. Without real assets, there is nothing to represent; and without real value, there is nothing to sustain.

The Role of Trust

Throughout human history, trust has been the single most expensive commodity in any economic system. To establish trust, our societies have relied entirely on massive, centralized institutions—banks, governments, legal systems, and corporate registries.

These centralized entities held the master ledgers of civilization. They verified who owned what, processed transactions, and resolved disputes. While this centralized model allowed modern commerce to scale, it also concentrated immense power, control, and fee-extraction capabilities in the hands of a select few intermediaries.

Tokenization introduces a completely different architecture of trust. It utilizes distributed ledger technology to establish a system where trust is secured by mathematics, immutable rules, and decentralized consensus.

This technology does not seek to violently replace traditional institutions overnight; rather, it fundamentally redefines how trust is distributed across society. When the ledger of ownership becomes transparent, mathematically verifiable, and accessible to everyone, the absolute power of centralized middle-men begins to evaporate, returning systemic leverage to the individuals who actually generate the value.

Where This Is Already Happening

We must understand that this shift is not a futuristic prediction or a distant science-fiction scenario. It is actively unfolding right now.

Across every major financial capital on earth, the most sophisticated asset managers, global banks, technology companies, and sovereign governments are actively building the infrastructure for this new ledger system. Trillions of dollars of commercial real estate, private equity, government treasuries, and gold are currently being tokenized.

They are doing this quietly, systematically, and with absolute focus, because they understand that the entity that controls the digital ledgers of the next era will control the flow of global commerce. It is not an ideological movement; it is an infrastructure race.

Traditional Assets, Creative Output, and Community Capital → Tokenization Ledger → Global Yield

What This Means for You

At this historic inflection point, the critical question is no longer whether tokenization is going to happen. The question is: *do you possess the cognitive literacy required to understand it while it is still in its infancy?*

Understanding this technology is the only way to establish strategic positioning for your life. You do not need to become a blockchain developer, and you do not need to make reckless financial moves. But you must cultivate a baseline level of awareness.

As the world's asset systems continue to evolve, participation will increasingly favor those who understand how the new ledgers are structured, while systematically leaving behind those who remain blind to the design.

It all comes back to the same three core pillars: **Assets, Ownership, and Participation**. These are not new concepts; they are the ancient, unchanging rules of human survival. But the way they are being connected, represented, and accessed is undergoing the most significant evolution since the invention of the joint-stock corporation.

Once you see this structural pattern clearly, the ultimate question of your life shifts: *How will this technology reshape who has access to ownership—and who will be permanently left on the outside looking in?*

The answer to that question will not be decided by computer code alone. It will be carved out by design, policy, education, and collective awareness. It will be decided by the people who have the courage to step forward, analyze the structure, and build the pathways that connect everyday communities directly to the real-world assets they generate.

The Bridge Forward

We are no longer simply discussing the abstract mechanics of technology; we are mapping the very systems that will connect human beings to value in the next era. When the capability to represent ownership expands, the implications transcend Wall Street and global financial markets—they flow directly into our local neighborhoods, our community programs, and our daily lives.

If ownership can be structured differently, then participation can finally be expanded to include those who have been locked out for generations. The future does not have to be a repeating loop of the past's inequalities. But this transformation depends entirely on one critical variable: not the speed of the technology, but the depth of human understanding.

Once the systems of the world learn how to represent ownership digitally, they will never go backward. What can be represented will be structured; what can be structured will be measured; and what can be measured will ultimately be controlled. The shift is not merely technological—it is deeply, permanently structural.

With this clarity, the conversation around ownership changes forever. The playground of the wealthy is becoming the battleground for the economic survival of everyone. And the path forward is no longer a matter of choice; it is a matter of design.

Chapter 13
THE TOKENIZATION OF EVERYTHING

Global Infrastructure. Human Consequences.

The Invisible Current

This is not a distant, speculative idea designed for academic debate. This is not some fragile concept forming quietly at the extreme edges of the global economy, waiting for some far-off date to prove its relevance. This is an active, hyper-accelerated evolution that is already in motion. It is unfolding beneath the noisy surface of daily life, away from the sensationalized headlines of modern media. It is happening within the quiet architecture of our most powerful institutions, across critical supply chains, inside global industries, and throughout sovereign nations. A completely new infrastructure layer is being systematically constructed—a programmable network designed to identify, organize, verify, and move value across the globe with absolute transparency. This ledger system is silently redefining who has access to property, how transactions settle, and how communities participate in the wealth they help create. Whether this shift is recognized by the public or completely ignored, it has already begun to fundamentally reshape the coordinates of the global economy. This is the dawn of what is rapidly becoming the tokenization of everything.

The Infrastructure You Cannot See—Yet

What makes this historic inflection point different from any paradigm shift of the past is not merely the brilliance of the underlying idea. The conceptual theory of digitized, programmable ownership is not brand new; it has been discussed, written about, and tested in early experimental environments for years. What makes this exact moment entirely different is adoption. We have officially transitioned from the era of curious observation to the era of massive, institutional construction. Real-world asset tokenization has quietly expanded from an experimental concept into a multi-billion-dollar global infrastructure, and its rate of growth is accelerating at an exponential scale.

Legacy Systems → Tokenized Rails

Legacy Systems	Tokenized Rails
Paper Deeds & Silos	Immutable Digital Ledgers
T+2 Settlement Friction	Real-Time Global Settlements
Restricted Institutional Access	Democratized Fractional Ownership

Yet, the raw mathematical numbers are not the primary signal we must focus on. The true signal is the level and caliber of participation. It is about who is actively building, who is aligning their capital, and who is quietly preparing their infrastructure for the new loop. The largest financial institutions, asset managers, and sovereign central banks on earth are no longer sitting on the sidelines watching this transition unfold. They are positioning themselves with absolute strategic focus.

Multi-trillion-dollar asset management firms are launching natively tokenized funds. Global banking conglomerates are systematically developing private blockchain networks where institutional capital moves on programmable, cryptographic rails. Traditional securities exchanges are rapidly re-engineering their platforms to prepare for a hyper-liquid future where global trades settle instantly and continuously.

Even money itself—the very lifeblood of human transaction—is beginning to move differently. When the underlying foundation of currency undergoes a structural evolution, everything built on top of that currency must inevitably shift as well. This is not a violent, overnight replacement of the traditional financial system. This is a profound, systemic convergence. The legacy systems we have trusted for generations are merging with decentralized digital networks to operate on a single, unified ledger system. Once this technical foundation is fully locked into place, the question of whether this technology matters will become entirely obsolete. The only question that will determine your survival is whether you possess the literacy to navigate it, or whether you remain blind to its design.

A System Built for a Different Time

To comprehend the sheer necessity of this transition, we must step back and examine the structural cracks in our existing models. The modern economy is hyper-digital, deeply connected, and operating at the speed of light. Yet, the legal and financial systems we rely on to record, verify, and transfer ownership are archaic relics of a bygone era. They are built upon paper documents, manual signatures, slow-moving administrative databases, and fragmented registries that cannot communicate with one another. We live in a world where a text message can travel across the globe in milliseconds and data is processed instantly, yet proving your legal ownership of a piece of land or a share of a business still requires days, weeks, and an army of expensive, centralized intermediaries.

In the friction of that administrative gap, a massive structural evolution has taken root. This movement does not seek to artificially create new value out of thin air; rather, it seeks to reorganize existing real-world value. It is a new method of representing what is already real—tracking it with mathematical precision, distributing it fluidly, and making it accessible to those who were historically locked out of the system. This transformation is not limited to a single sector or a specific asset class. It is systematically touching everything: commercial real estate, private operating businesses, global supply chains, national debt instruments, intellectual property, creative masterworks, and the collective capital of local communities.

What This Makes Possible

If this emerging digital infrastructure is designed with conscious clarity and moral intent, it holds the power to create unprecedented pathways for human advancement. It can open the gates of systemic access, transforming high-value assets that were once restricted to the ultra-wealthy into divisible, fractional units that anyone can participate in. It can introduce absolute, radical transparency to systems that have historically operated in the dark, making the flows of capital and the structures of ownership visible and verifiable on public ledgers.

Furthermore, this technology can eliminate the massive administrative friction that drains value from transactions, creating an incredibly smooth, efficient, and aligned economy. At its absolute highest potential, tokenization allows everyday people to transition from being passive consumers trapped inside systems to active, sovereign participants who understand, own, and help shape the very networks they depend on for survival.

Systemic Attribute	The Sovereign Path (The Light)	The Centralized Path (The Shadow)
Access	Open, permissionless fractional participation.	Strict gatekeeping and systemic exclusion.
Transparency	Publicly verifiable, mathematical truth.	Total behavioral surveillance and trackable control.
Velocity of Capital	Instant global settlement and fluid local yields.	Accelerated wealth concentration and automated extraction.

Every System Has a Shadow

However, we must remain fiercely realistic as we stand at this historical crossroads. Every major technological breakthrough carries dual potential. The same systemic architecture that has the capacity to expand human freedom and democratize wealth can also be engineered to restrict access and centralize power. The identical infrastructure that provides radical transparency can, if centralized in the hands of a few, be converted into an unprecedented apparatus of financial surveillance. The same computational efficiency that speeds up the flow of capital can easily accelerate the concentration of wealth, widening the Ownership Gap into a permanent, insurmountable canyon.

This is not a warning designed to breed fear, panic, or withdrawal. It is a call for deep, analytical awareness. Technology itself is completely neutral; it does not possess a moral compass or determine human outcomes on its own. The ultimate trajectory of this transition will be decided by design, by intent, and by the values of the human beings who build the code.

The Balance That Must Be Held

Navigating this transition successfully does not require emotional panic. Fear is a highly destructive, reactive state of mind; it causes people to rush blindly into speculative trends, distorts their analytical clarity, and forces them into defensive, ideological corners. True positioning requires calm, disciplined awareness. Awareness possesses the capacity to pause, step back, study the patterns, and make intentional, strategic choices.

What is being built right now is not a minor technological upgrade. It is the core financial and property infrastructure that will dictate how value is accessed, how ownership is structured, and how opportunity is distributed across society for generations to come. The system itself will not decide whether this power is used to liberate or control. The people within the system must decide. Every builder, every creator, every entrepreneur, and every local leader must actively step into this conversation today to ensure that the code we write reflects human dignity, trust, and shared prosperity.

Chapter 14
THE FUTURE BEING BUILT IN REAL TIME

Abundance, Awareness, and Human Possibility

A Generational Inflexion Point

We are currently living through a profound, historic transition that future generations of historians, economists, and sociologists will study with intense fascination. We are not merely witnessing a standard cultural cycle, a passing political shift, or a temporary economic downturn. We are standing directly inside a permanent, exponential transformation of the human experience. It is a fundamental shift in how the world operates, how value is constructed, and how human beings participate in the systemic networks that dictate their daily survival.

This is not a futuristic prediction, a theoretical scenario, or a distant science-fiction blueprint. This is our immediate, structural reality. The future of human civilization is not waiting to be imagined; it is actively being built right now. It is being codified in software architectures, integrated into modern financial policy, executed in global markets, and lived out in the daily decisions of ordinary communities. Because these foundational systems are being built in real time, we have been granted a narrow, highly critical window of opportunity. This is the precise moment where active awareness can still be converted into structural positioning, and where the decisions we make today will permanently dictate how our children experience tomorrow.

A World Moving Toward Abundance

If we train our eyes to look past the superficial noise of modern media, we can easily see the powerful structural signals of a new era. Artificial intelligence is expanding its cognitive capabilities at an exponential rate, advanced automation is rapidly streamlining complex industrial processes, and robotics are scaling with remarkable physical precision across every sector of global production. Technology is becoming capable of producing a massive volume of goods, services, and energy—faster, cheaper, and at a scale that was completely unimaginable to previous generations. This massive technological leap introduces a historic possibility: the creation of a world where structural scarcity is systematically dismantled.

The Pathway to Autonomous Revenue

Path	Flow
Old Labor Paradigm	Labor → Wages → Consumer Survival
New Asset Paradigm	Capital → Asset → Autonomous Yield

Imagine a world where the baseline physical needs of human survival are incredibly simple and inexpensive to meet. Think of a society where food production is completely optimized, where local manufacturing is hyper-efficient, and where automated networks dramatically drive down the cost of daily living. In such a world, human beings would finally be liberated from the exhausting, daily struggle for basic physical survival. We would no longer be defined solely by the labor we perform to keep the wheels turning. Instead, we would be structurally free to redirect our finite cognitive energy toward creativity, deep innovation, local community building, and personal purpose. At its highest potential, this is the magnificent promise of a technology-driven age of abundance.

Humanity and Technology—Together

In this optimized future, technology does not exist to replace, displace, or diminish humanity. Rather, it exists to support, elevate, and empower it. It is designed to remove the brutal, repetitive physical friction of survival, automate systemic inefficiencies, and create the physical and mental space required for human beings to focus on what matters most. Human beings bring the essential elements that can never be replicated by silicon or algorithms: creative vision, emotional depth, moral guidance, and deep, empathetic connection. Technology brings computational speed, industrial scale, and mathematical precision.

When these two forces are brought into harmonious alignment, they create an incredibly powerful partnership. They establish a cooperative system where the systemic crises that have plagued human society for centuries—such as localized poverty, inadequate access to clean resources, and educational inequality—can finally be solved at scale. The physical foundation for this abundant future is already being laid down as we speak.

A Future Built on Values

However, we must remain deeply grounded: technology alone is entirely insufficient to secure a prosperous future. The true foundation of an abundant world cannot be built on computer code or physical machinery alone. It must be rooted in unyielding human values:

- **Trust:** The absolute certainty that our digital networks and ownership ledgers are designed with mathematical fairness, absolute transparency, and decentralized verification.
- **Love:** The conscious, ethical commitment to ensure that no individual, family, or local community is discarded or forgotten in the pursuit of physical optimization.
- **Freedom:** The preservation of individual sovereignty, ensuring that every human being retains the autonomous capacity to think, choose, build, and own.

Without these foundational human values, even the most technologically advanced system can easily degenerate into a cold, hyper-efficient machine of control. But when these values are actively engineered into the very code of our emerging networks, technology is transformed into the ultimate tool for shared human progress.

The Role of Ownership in an Abundant World

As our productive systems become increasingly automated and efficient, an urgent, systemic question arises: *who ultimately owns and participates in the immense value being generated?* Abundance at the level of production does not automatically guarantee access at the level of consumption. If the engines of technology produce an endless flow of wealth, but the legal and structural ownership of those engines remains highly concentrated in the hands of a select few, the Ownership Gap will simply expand.

This is why the concept of ownership becomes more critical in an abundant world, not less. Ownership is not about the greedy accumulation of material goods. It is about systemic participation. It is the direct connection that links an individual or a community to the wealth-generating engines of society. It ensures that the fruits of technological progress are not merely distributed as temporary, state-managed charity, but are securely owned by the people themselves.

The Power of Local Communities

This abundant future is not a top-down design that will be delivered to us by elite boardrooms, Silicon Valley laboratories, or global financial institutions. It must be built from the ground up, starting directly within our local communities. It must take form in places like Buffalo, New York, in neighborhoods across Atlanta, Georgia, and in every local environment where ordinary people create real, physical value every single day.

When local families, independent entrepreneurs, local artists, and community leaders combine their hard work with systemic awareness and ownership literacy, a profound shift occurs. They transition from passive, vulnerable participants to active, sovereign builders. They stop merely reacting to the economic shocks of the world and begin intentionally positioning themselves within the new infrastructure. This is how we construct a future where technological progress and human prosperity move in perfect, lasting alignment.

Chapter 15
HOW OWNERSHIP COULD EVOLVE FOR SMALL BUSINESS

From Work to Asset. From Asset to Ownership.

The Ground Floor of Evolution

If the very definition of ownership is undergoing a profound, systemic evolution, then this conversation cannot remain confined to global financial capitals, institutional boardrooms, or abstract technology conferences. It must occur closer to home. It must take shape on our local main streets, inside our neighborhood barbershops, across our kitchen tables, and on our physical job sites. It must manifest in the daily lives of the hardworking men and women who possess the courage to build something real out of nothing.

The local businesses that sustain our neighborhoods are not separate from this global technological transition; they are its absolute center. Once you perceive this connection, your relationship with your work changes permanently. You stop viewing your business as a temporary job that merely generates current income, and you begin analyzing it as a powerful, structured real-world asset that can be positioned strategically to capture long-term, compounding value.

What We See vs. What Is Truly There

If you walk through any neighborhood with a trained, observant eye, the ordinary movement of daily life begins to reveal its true, structural design. You see the multi-generational family restaurant serving customers, the local hair salon acting as a hub of culture and deep neighborhood conversation, the independent retail store lining the street, and the construction crew pouring concrete for a new building.

To the passive observer, this looks like mere physical activity. But when your perception shifts, you realize you are looking at a dense network of powerful, real-world assets. These local businesses are the primary economic engines of our society. They generate stable local employment, circulate capital within the neighborhood, build deep trust, and establish the cultural identity of our cities. Yet, despite the massive, undeniable value they create every day, they have historically been structurally barred from fully owning and scaling that value.

What He Built With His Hands

I find myself thinking about this reality often, especially on the days when I step away from my writing desk to work alongside a friend of mine who owns an independent construction company in Western New York. A few times each month, I join his crew on various job sites. Some days we are executing complex residential renovations; other days we are doing commercial fit-outs; and on some of the best days, we are doing heavy demolition—swinging sledgehammers, tearing down old, decayed plaster walls, clearing out what was, and making physical room for what is coming next.

The Value Conversion Pathway

Raw Effort	Trust & Reputation	Digital Equity
Physical Labor	Client Relationships	Tokenized Assets
Active Working	Brand Capital	Multi-Gen Wealth

I do not coach formal collegiate athletics as much as I used to, so getting out on a raw demolition site with him gives me a chance to physically push myself, sweat, and get after it with the same competitive energy I carried on the football field. There is a beautiful, undeniable honesty in manual labor. There are no shortcuts, no performative acts, and no hiding behind comfortable words. You look directly at the raw material in front of you, apply focused, physical energy, and piece by piece, you reshape it into a completely new structure.

We first met years ago through the game of football. We coached together at the junior college level, and later guided young athletes at the high school level. Because of those intense, shared seasons on the turf, his entire family knows me well. To his children, I am not an author, a speaker, or an analyst; I am simply "Coach Teddy."

Recently, we were working on a beautiful house he had purchased in Clarence, New York—a property he is personally renovating and transforming into a permanent home for his family. As we worked, the steady rhythm of our labor was accompanied by the deep, running conversations we have shared for years: business strategy, family, philosophy, and where the world is going.

From time to time, I would bring up the concepts of real-world assets and the emergence of tokenization. I did not weigh the conversation down with dry, technical jargon or complex computer science definitions. Instead, I focused on the core, human idea: how the global systems surrounding property rights are changing, and what that structural shift means for builders, operators, and small business owners like him—men who have spent their entire adult lives pouring their physical souls into creating something tangible.

Sometimes he would capture a piece of the vision. Other times, he would pause, wipe the sweat and plaster dust from his brow, smile, and give a response that perfectly captures how the vast majority of hardworking people feel when they first hear about this technology: "*It's definitely interesting, Teddy*," he said one afternoon. "*It's probably something we should take the time to learn more about instead of being afraid of. I guess I'll understand it a lot better once I finally read your book.*" We laughed, shook off the dust, and went straight back to swinging our hammers.

But during that same project, a simple moment unfolded that changed my perspective forever. His young son walked onto the job site, looked up through the framing studs, and said, "*Hey, Coach Teddy.*"

In that precise moment, the abstract concepts of asset structure and legacy became instantly concrete. I was no longer just looking at my friend as a local contractor executing a renovation project. I was looking at the immense, structured monument of his life's work. I saw the construction company he had built with his own hands, the hundreds of successful projects completed across Western New York, and the flawless reputation earned job by job, client by client, through decades of sweat, honesty, and integrity.

I thought about the deep relationships of trust he had established throughout the community. I thought about the phone calls he receives daily, not because of expensive marketing campaigns, but because people know that when he puts his name on a contract, he shows up and does the job with absolute excellence. I thought about the grassroots youth ministry he is building, pouring his personal capital and time into mentoring young people the same way we once did on the sidelines of the football field.

Then, the ultimate truth of the Ownership Pathway™ revealed itself. None of this was just temporary economic activity. None of this was random. This was massive, highly concentrated value—structured through relentless human effort, built through unwavering consistency, and sustained through decades of community trust.

The company, the relationships, the reputation, the body of work, and the positive, cascading impact on the lives of local youth—all of these were not merely tasks he was performing to earn a living. These were real-world assets. They held immense, compounding value that had the capacity to outlast his physical labor, support his children across generations, and establish a permanent legacy.

The Challenge of the Legacy Loop

Yet, despite the massive value that independent builders and small business owners generate every single day, they are trapped inside one of the most restrictive loops in our entire economic system. They create the physical and cultural wealth of our nation, but they are seldom taught how to legally or technologically structure that wealth so that it can grow independently of their physical labor.

Historically, when a local builder, a restaurant owner, or a service provider stops working, the income stream stops instantly. The value they created remains highly localized and illiquid, unable to be leveraged, scaled, or transferred efficiently to the next generation. They have been given the tools to work, but they have never been shown the playbook to own.

The Small Business Asset Transition Playbook

To bridge this painful gap, we must actively translate the conceptual evolution of ownership into a practical, repeatable operational framework that small business owners can execute today.

Step	Focus Area	Operational Objective	Core Strategic Question
1	**Strategic Perception**	Transition from viewing the business as an income stream to analyzing it as a stand-alone asset.	*Does this business generate intrinsic, structural value independent of my daily physical labor?*
2	**Value Quantification**	Identify and document intangible assets, including customer trust, brand equity, and community reputation.	*How can we measure, protect, and legally structure our reputation capital as a tangible asset?*
3	**Systems Architecture**	Standardize, document, and automate	*Are our core operational loops automated and*

Step	Focus Area	Operational Objective	Core Strategic Question
		operational processes to prepare the business for scalable growth.	*repeatable without my manual intervention?*
4	**Directional Awareness**	Align the business model with macro shifts, including digital ledger technology and automated systems.	*How will evolving regulatory frameworks and tokenization platforms impact our specific industry?*
5	**Strategic Positioning**	Build deep connections with forward-thinking builders and secure early positions near digital value hubs.	*Where can we align our services with emerging local asset networks to build strategic leverage?*
6	**Bedrock Protection**	Shield the core assets of the business through unyielding integrity and consistent value delivery.	*Are we actively protecting our community relationships as our primary source of systemic yield?*

The Bridge to What is Next

This is not about making the life of a small business owner more complicated, stressful, or administrative. It is about helping them clearly recognize the true, monumental nature of what they have already built. They are not merely workers exchanging time for money; they are the active creators of our society's foundational assets.

Once this awareness takes root within our neighborhoods, the conversation around ownership naturally expands. It moves seamlessly into other high-visibility arenas where human performance, cultural influence, and immense value creation meet. It flows directly into the world of sports—an industry where for generations, the individuals who created the ultimate value on the field were treated as temporary performers, while the structural equity of the system remained locked behind institutional walls. But the field is changing, the players are waking up, and a new paradigm of ownership is about to be written.

THE PLAYBOOK

A Framework for Small Business Ownership

If ownership is evolving—then small businesses must evolve with it.
Not by abandoning what works—
but by understanding what they truly are.

1 See it Clearly

A business is not just income.
It is an asset.

2 Define The Value

- Revenue.
- Relationships. Trust.

3 Build Structure

- Clarity before scale.

4 Stay Aware

- Not chasing trends—understanding direction.

5 Position Early

- Because positioning shapes outcomes.

6 Protect The Foundation

- Trust.
- Consistency.
- Real value.

The Playbook for Small Business

Chapter 16
FROM COMMUNITIES TO ATHLETES

Understanding Ownership in Motion

The Arena of Value and Exploitation

As we trace the structural evolution of property and equity from physical soil to programmable code, our conversation must inevitably shift into a highly visible arena where immense human value has always vibrated at a fever pitch, yet where true ownership has historically been systematically withheld. We must step onto the field of sports.

For generations, athletes have created an almost unimaginable volume of economic and cultural value. They fill massive metropolitan stadiums with crackling, electric energy; they drive global broadcast rights deals worth billions of dollars; and their names and likenesses sell merchandise in retail markets across every continent on earth. Entire cities and local regions routinely rally around their favorite teams and players, utilizing sports not merely as a temporary weekend escape, but as a core pillar of their regional identity, pride, and shared community connection. The raw value generated by these performers has never been the problem; it has always been present, compounding, and highly visible.

Yet, when we look beneath the surface of this massive cultural engine, we find the exact same structural pattern that governs every other legacy system in our society: value is relentlessly created by the many, but the long-term equity generated by that value is captured exclusively by a select few. For the vast majority of athletic history, the very performers who put their physical bodies on the line to fuel this multi-billion-dollar apparatus possessed virtually zero structural control over the commercial value connected to their own names, images, and likenesses. They were treated as high-performing, temporary labor—glorified performers on a closed stage—while the permanent equity of the system remained locked behind institutional walls.

NIL Is Not What It Seems

For decades, this exploitative loop was defended as an immutable rule of amateurism and institutional tradition. Athletes poured their blood, sweat, and youth into their sports, only to see their institutions grow wealthy while they were legally barred from capturing even a fraction of the value they generated. This systemic barrier was not a reflection of a lack of talent or influence on the part of the players; rather, the system itself was intentionally engineered to deny them the rights of ownership.

However, we have officially entered a period of historic disruption. With the arrival of Name, Image, and Likeness (NIL) regulations, a massive structural door has been kicked wide open. For the very first time in modern sports history, athletes at the collegiate level and beyond are legally empowered to participate directly in the commercial value of their personal identities.

Yet, this is precisely where the vast majority of the public, the media, and even the athletes themselves completely misunderstand the moment. To the casual observer, NIL is viewed through a highly transactional, short-term lens. It is discussed in sports talk radio and online forums as a collection of quick cash deals, lucrative endorsement contracts, and temporary corporate sponsorships designed to pay young athletes in the moment.

But from a deeper, structural perspective, NIL is not merely a new monetization channel. It is a profound, historic signal that the boundaries of ownership are actively expanding into spaces where they were once fiercely restricted. It is a legal and cultural recognition that identity itself—your name, your image, your reputation, and your unique presence—is no longer just an abstract human quality. It is a structured, highly valuable real-world asset.

The Value Conversion Pathway

Raw Talent	**Active Performance**	**Digital NIL Assets**
Physical Labor	On-Field Conviction	Tokenized Brands
Transient Output	Relationship Capital	Multi-Gen Equity

The Divide Within the Locker Room

As this new landscape takes shape, a deep, invisible division is beginning to form within locker rooms across the country. It is a divide that has nothing to do with athletic talent, physical speed, or game-time execution, but is entirely rooted in cognitive perspective and alignment. On one side of this divide stand the athletes who perceive NIL solely as a temporary transactional opportunity. They view the new regulations through the eyes of a consumer, treating their incoming capital as a quick check to be immediately spent on depreciating material goods. They are focused on the immediate payout, and when their athletic careers inevitably come to a close, the financial opportunity vanishes alongside their physical playing time.

On the other side of the divide stand a rare, focused group of athletes who recognize that NIL is not the destination, but the ultimate gateway to strategic positioning. They understand that their playing years represent a highly compressed, high-visibility window where they can leverage active attention to build a permanent, structured infrastructure of wealth. They do not view themselves merely as transient players on a roster; they analyze themselves as the sovereign founders of a digital corporate enterprise. They use their current platform to secure equity, establish independent business structures, and align themselves with long-term assets that will continue to yield value decades after they have played their final down of football. They understand that in a rapidly shifting economy, the ultimate game is not played on the turf, but on the master ledgers of ownership.

A Hard Truth from the Sidelines

As a veteran coach who has spent over twenty years on the sidelines, in the film rooms, and in the lives of young athletes, I have had to confront some highly uncomfortable truths about my own legacy. Throughout my career, I have had the profound privilege of coaching, mentoring, and developing young men of immense character, discipline, and relentless drive. I have watched with pride as some of them earned coveted Division I scholarships, others stepped onto the elite fields of the NFL, and some transitioned into coaching to pass the torch to the next generation. I am deeply proud of every single one of them.

But when I look back at my decades of coaching with absolute honesty, I cannot run away from a glaring structural omission: I never taught them ownership.

We coached them to be tough. We coached them to be disciplined, to lead under pressure, and to sacrifice their individual desires for the collective goals of the team. We prepared them meticulously to compete inside the physical lines of the game. But we left them completely blind and unequipped to navigate the ruthless economic systems waiting for them on the other side of those lines. We trained them perfectly for high-value labor, but we never showed them the playbook for asset preservation.

I have watched too many young men step through the doors of elite athletic opportunity, only to find themselves completely lost in the administrative and financial currents of the real world because they lacked structural awareness. Talent can open the door of a massive stadium, but without ownership literacy, that door will eventually slam shut, leaving you with nothing but fading memories and a worn-out uniform.

Watching the Shift as a Father

Today, this realization has transitioned from a professional concern into an intensely personal reality. I am no longer just watching this economic shift unfold from the coaching sidelines; I am living it through the eyes of a father. My own son is currently navigating the highly complex, high-visibility world of Division I college football. As I watch him step onto the field each week, I see him operating in an athletic and financial landscape that simply did not exist when I was competing in the 1990s. He is surrounded by non-stop brand conversations, complex

NIL contracts, social media monetization, and global visibility at scale.

While this new era provides unprecedented financial opportunity, it also introduces a terrifying level of exposure. Young athletes are being thrust into sophisticated corporate environments and handed highly complex financial instruments earlier than ever before in human history—often without possessing even a baseline level of structure or education to navigate them safely. In this high-stakes landscape, the risk of exploitation is higher than it has ever been.

Access to quick capital without structural understanding does not produce long-term freedom; it simply creates a louder, more temporary form of consumption. Without a clear personal playbook, young athletes are highly vulnerable to signing away their ultimate leverage to predatory intermediaries who understand the system far better than they do. This is why we must immediately bridge the gap between physical performance and economic literacy, ensuring that our young people are equipped to capture, structure, and own the incredible value they create.

The NIL Conversation Is Bigger Than Sports

To fully comprehend the magnitude of this transition, we must lift our eyes beyond the athletic fields and locker rooms. The concept of Name, Image, and Likeness may have found its primary legal battlefield in college sports, but the underlying principle of NIL belongs to the entire modern world. This was never just a sports debate; it is a foundational argument about the legal and technological ownership of human identity.

Athletes are simply the most visible pioneers of a shift that is currently touching millions of ordinary people across global digital networks. Every single day, on platforms like YouTube, Instagram, TikTok, and Substack, creators are generating immense, measurable value through the monetization of their personal identities. They are building massive global audiences, shaping cultural conversations, directing attention flows, and driving multi-million-dollar economic ecosystems.

Yet, the vast majority of these digital creators are operating under the exact same NIL reality as collegiate athletes—they simply lack the language to recognize it. Your unique voice, your creative presence, your intellectual reputation, and your consistent attention-capture are not just casual forms of self-expression. In a hyper-connected, digital economy, your identity is your primary commercial asset.

Unfortunately, most creators fall into the exact same structural trap as the transactional athlete. They focus exclusively on immediate monetization—chasing quick brand deals, temporary platform payouts, and volatile algorithmic ad revenue—without ever building an independent infrastructure of ownership. They act as digital sharecroppers, building massive value on rented platform land that can be reclaimed, censored, or demonetized overnight by a single algorithmic shift. True sovereignty requires a transition from passive platform consumption to structured asset creation.

The Creator NIL Playbook

To help creators navigate this digital landscape, we must actively translate the principles of athletic identity ownership into a practical, repeatable framework.

Stage	Focus Area	Actionable Strategy	Sovereignty Objective
1	**Recognize Your Value**	Transition from viewing content creation as a hobby to analyzing it as a consistent asset engine.	Audit your audience retention and recognize the baseline economic value of your attention-capture.
2	**Protect Your Identity**	Treat your name, image, and voice as corporate intellectual property with strict legal boundaries.	Secure proper trademarks and incorporate your digital identity within a structured legal entity.
3	**Sovereign Infrastructure**	Move your audience away from rented third-party platforms and onto independent, owned digital channels.	Build direct-to-consumer infrastructure, including owned websites, email databases, and direct networks.
4	**Asset Securitization**	Convert transient creative output into structured, licensable, and programmable intellectual property.	Package your digital content into distinct, reusable educational products, media libraries, or programmatic assets.
5	**Generational Leverage**	Focus on building deep trust and long-term community relationships rather than chasing viral algorithm metrics.	Leverage your platform to acquire real-world equity, establishing a compounding legacy that outlasts your current labor.

The Athlete Ownership Playbook

Just as digital creators must structure their platforms, athletes must utilize a clear, disciplined playbook to convert their high-visibility physical careers into permanent financial leverage.

Playbook Pillar	Core Concept	Operational Shift	Ultimate Outcome
1. Asset Identification	You are not merely a player on a roster; you are a sovereign corporate enterprise.	Transition from a laborer's mindset to analyzing your career through the lens of enterprise equity.	Establish a clear, professional brand identity that commands value independent of your roster spot.
2. Talent vs. Ownership	Physical talent gets you in the door, but structural ownership dictates what you keep.	Focus on acquiring and retaining asset rights rather than simply working for a temporary salary.	Build a permanent financial foundation that continues to grow long after your athletic career concludes.
3. Beyond the Deal	NIL is not the destination; it is the starting point for long-term equity acquisition.	Pivot away from short-term transactional checks toward acquiring direct equity in the brands you support.	Secure a compounding ownership stake in the economic systems you help scale with your influence.
4. Structural Protection	Value created must be legally and administratively structured to be fully captured.	Build a trusted, highly vetted advisory team consisting of legal, financial, and strategic experts.	Shield your active earnings from predatory fee-extraction and administrative leaks.

Playbook Pillar	Core Concept	Operational Shift	Ultimate Outcome
5. Simultaneous Construction	The timeline of the game is highly compressed; the timeline of ownership is infinite.	Actively build your business networks, relationships, and assets while you are still playing the game.	Smooth the transition out of professional sports by having an active asset infrastructure already in place.
6. Strategic Positioning	Avoid chasing volatile financial trends; position yourself near stable, sovereign value hubs.	Focus your attention on acquiring real-world assets, including commercial real estate and cash-flowing businesses.	Anchor your legacy in tangible wealth engines that are completely immune to athletic injury.

The Trajectory of the System

The macro pattern of human progress is incredibly consistent: first comes the spark of awareness, followed by strategic positioning, active participation, and ultimately, true structural ownership. Today, we are witnessing a massive collective awakening within the world of sports. Athletes are slowly realizing that the old rules of the game were engineered to exploit their labor while denying them a stake in the infrastructure. They are beginning to participate in the wealth of the system, but only a rare, focused few will successfully transition into true structural owners of the world around them. This cognitive difference will permanently dictate life outcomes for the next generation of builders.

No matter how advanced or digitized our financial systems become, the bedrock rules of sovereignty never change. You must understand your value, you must protect your identity, and you must master the structural mechanics of the ledger systems you operate within. When elite talent is paired with conscious, disciplined ownership literacy, the transient performer is transformed into a sovereign creator. They stop merely playing the game, and they begin to write the rules.

This economic transformation carries a profound impact that extends far beyond individual financial success. When an athlete or a creator from an underserved community successfully bridges the gap into true structural ownership, they do not just change their own personal trajectory. They establish a permanent, cash-flowing monument of wealth that can support families, fund local businesses, and provide real opportunity for their entire neighborhood across multiple generations. This is how we systematically dismantle the Ownership Gap—not through top-down institutional charity, but through a grassroots, compounding movement of economic sovereignty.

The Next Arena of Sound

Once you train your mind to identify this structural pattern, you begin to recognize its presence in every single corner of our cultural and economic landscape. We have traced its design across local neighborhoods, small business operations, and elite athletic fields. Now, our conversation must move into another high-visibility arena where human creativity, emotion, and culture meet to generate astronomical amounts of global wealth—yet where the creators themselves have historically been subjected to some of the most predatory, exploitative structures in human history. We must step into the dark, brilliant, and complex world of the music industry.

ATHLETE OWNERSHIP PLAYBOOK

From Performance → Participation

Not just in the game—
In the game beyond the game.

1 YOU ARE AN ASSET

Not just a player.

- A brand.
- A platform.
- A source of value.

Your performance creates attention.
Your identity builds connection.
Your consistency builds trust.

2 TALENT IS NOT OWNERSHIP

Talent gets you on the field.
Ownership determines what you keep.

Because the game ends.
Ownership does not.

3 NIL IS THE BEGINNING

Not the finish line.

- Ownership moves beyond endorsements—into:
  - Equity
  - Structure
  - Long-Term Participation

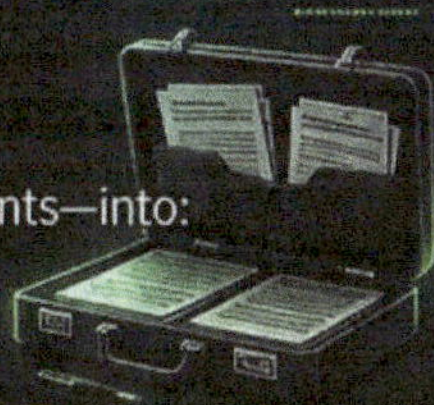

4 BUILD YOUR INFRASTRUCTURE

Without structure—
value is created... but not captured.

You need:
- Clarity
- Advisors
- Legal Understanding

5 BUILD WHILE YOU PLAY

The game has a timeline.
Ownership doesn't.

So the real question becomes:
What are you building while you are playing?

6 LEGACY IS THE GOAL

Impact outlives your jersey.
Legacy outlives your name.

Build what helps others rise
long after you've moved on.

Chapter 17
THE MUSIC INDUSTRY AND OWNERSHIP

Creativity, Rights, and Long-Term Value

The Studio at Midnight

The control room of the recording studio was completely dark, save for the soft, warm glow of a single lamp in the corner and the flickering green and red LEDs of the mixing console. The air carried the faint, familiar scent of old dust, warm electronics, and highly concentrated creative energy.

Through the monitors, a heavy, stripped-down beat was playing on a continuous loop—not loud, but vibrating with a steady, rhythmic bass that sat deep in your chest. The artist sat quietly on the couch, head nodding in perfect, silent alignment with the tempo, legal pad balanced on his knee, pen hovering over the page as words, melodies, and complex rhythmic structures began to take physical form in real time.

This was another late-night session extending long past midnight—the kind of invisible, grueling work that the public never gets to witness. There were no flashing cameras, no cheering audiences, and no roaring applause in this room. There was only the quiet, relentless, and sacred act of pure human creation.

Within this silent space, hours of focused mental and spiritual energy were slowly being condensed into a single recorded track that had the capacity to travel across the globe and last forever. This is where music begins. It does not originate in sterile legal contracts, institutional boardrooms, or corporate marketing meetings. It is born in these quiet, intimate moments of unfiltered human expression.

Music as Stored Human Capital

Throughout the history of human civilization, music has existed as one of our most potent, transcendent forms of expression. A single recorded song possesses the unique, almost magical capacity to travel vastly further than the physical human being who created it. It crosses oceans, breaks through cultural barriers, and echoes across multiple generations. It shapes our personal identities, carries our deepest collective emotions, and weaves itself seamlessly into the very fabric of our life stories.

Music acts as the sonic soundtrack to our most sacred moments—our celebrations, our heartbreaks, our triumphs, and our losses. It is a living testament to what it feels like to be alive.

But beneath this beautiful, emotional surface of sound, there is a cold, highly organized structural reality that most people never stop to analyze. Music is not merely an artistic expression; it is a highly lucrative, multi-dimensional financial asset.

Every single song that is conceptualized, recorded, and finalized is instantly converted into a distinct form of legally recognized intellectual property. This intellectual asset produces consistent, compounding streams of global value through diverse channels: digital streaming platforms, terrestrial and satellite radio, physical sales, synchronized licensing for film and television, and live public performances.

Yet, behind the beautiful melodies and the massive revenue flows lies a complex, layered architecture of legal rights and corporate structures that ultimately determines who captures that value, and who is left on the outside looking in.

The Hidden Structure of Sound

For the vast majority of aspiring artists, the physical act of creation is simple and intuitive, but the corporate legal structure surrounding their craft is an absolute, impenetrable maze. The music industry is a highly sophisticated, multi-tiered apparatus consisting of record labels, publishing companies, distribution networks, streaming platforms, and legal intermediaries. Every single transaction, release, and monetization event is governed by complex, multi-page contracts that are written in a dense, highly specialized legal language.

These contracts are the master ledgers of the creative world. They do not merely organize logistics; they decide with absolute legal finality who owns the master recordings, who controls the publishing rights, how the revenue flows are partitioned, and who ultimately gets paid over the long-term trajectory of a career.

Too often, young, hungry artists enter these highly leveraged agreements with intense, brilliant clarity about their musical craft, but absolute blindness regarding their intellectual property rights. They sign agreements in a rush of excitement and ambition, and by the time they finally develop the literacy to understand what they have signed, their ownership has already been permanently surrendered.

The Shadow of the Dream

This is the dark, silent reality of the music industry that is never celebrated in the glossy pages of entertainment magazines or on the high-profile stages of awards shows. The visible dream is brilliantly engineered and heavily promoted: fame, global recognition, massive wealth, and cultural influence.

But the structural machinery operating behind that dream is often ruthlessly extractive. We see legendary artists who have written and performed songs that have generated hundreds of millions of streams globally, yet who are struggling financially because they do not own their master recordings. We see brilliant songwriters who have penned timeless, generation-defining hits, yet who control zero percent of their publishing rights.

The Value Divergence Model

Path	Flow
Traditional Path	Labor → Streaming Advance → Debt Loop
Sovereign Path	Effort → Master Rights → Legacy Yield

The hardest, most tragic truth of this system is that some of the most culturally impactful and spiritually resonant music ever recorded is still generating massive, compounding financial value today—but not for the families of the creators. The surplus value of their genius continues to flow directly into the corporate vaults of institutional legacy systems that acquired their rights decades ago for a pittance. This is the Ownership Gap in its most painful, visible form.

The Master vs. The Publishing

To navigate this complex creative economy successfully, an artist must possess absolute, unyielding clarity regarding the two primary pillars of music ownership.

Ownership Pillar	Legal Definition	Asset Characteristics	Monetization Mechanics
Master Rights	The ownership of the specific physical or digital audio recording of a song.	Governs the literal sound file and the performance captured on the track.	Yields revenue from digital streaming, physical sales, and synchronization licensing in media.
Publishing Rights	The ownership of the underlying composition (lyrics, chords, and melody).	Governs the intellectual creation, the writing, and the musical arrangement.	Yields royalty streams from radio airplay, cover versions, public performances, and sheet music.

If an artist does not understand the structural difference between these two rights, they can easily make the catastrophic mistake of selling one or both of them away, effectively cutting themselves off from the lifetime yield of their creative output. You can perform to sold-out arenas across the globe, but if you do not own the master recording or control the publishing rights of your catalog, you are simply a highly paid employee of a corporation that owns the fruits of your labor.

A Beat Left on a Hard Drive

I find myself thinking about this reality often, especially during my deep conversations with a close friend of mine whom I consider a true brother in every sense of the word. He is one of the most intellectually sharp, culturally grounded individuals I have ever met—the kind of person who possesses the rare capacity to read a room instantly, see the hidden patterns of the world, and build deep, lasting connections with people from all walks of life.

Music has always been his primary passion. He is not just someone who casually enjoys songs; he is a dedicated writer, a student of rhythm, and a guardian of the culture. He would describe himself as a true, authentic MC—someone who carries immense respect for the historical foundations and spiritual weight of hip-hop.

Several years ago, a close friend and lodge brother of ours—a highly talented creator who was deeply involved in the local music scene—passed away suddenly. He left behind a rich, unfinished legacy, including a massive digital catalog of unreleased beats, musical arrangements, and production sessions.

Before he passed, he entrusted this entire catalog to my friend, leaving it in his hands to preserve and carry forward. This catalog was not just a collection of digital files on a hard drive; it was a physical monument of his life's work. It represented years of late-night studio sweat, focused human genius, and pieces of his very soul captured through sound.

My friend often speaks about the intense, sacred weight of this responsibility. He does not view this catalog as a commercial opportunity for personal profit; rather, he approaches it with the protective discipline of a trustee.

I will never forget when he looked at me during one of our late-night conversations and said, "*Teddy, my only goal is to protect this catalog so that one day, when his children are grown, these beats can provide a stable financial foundation for their lives."*

In that precise moment, the abstract concepts of music rights, intellectual property, and tokenization were instantly stripped of their dry, corporate skin. I was no longer looking at an archive of digital sound files. I was looking at a structured asset of generational legacy.

This beat catalog was a digital trust—a reservoir of stored value capable of outliving its physical creator, crossing the cold boundary of death, and returning to the physical world to support, protect, and educate his fatherless children. This is what true ownership makes possible. It is the transition from temporary transactional income to permanent generational legacy.

The Shift to Independent Sovereignty

Thankfully, the historical paradigm of absolute institutional exploitation is beginning to show deep structural fractures. A massive collective shift toward conscious sovereignty is currently taking root across the global creative landscape.

More artists than ever before are asking difficult, systemic questions. They are refusing to accept the predatory advance contracts of the past, choosing instead to own their master recordings, retain absolute control of their publishing rights, and leverage modern digital networks to distribute their music independently.

This movement is not a chaotic rebellion against the industry; it is a logical, disciplined alignment with structural reality. Creators are finally waking up to the realization that their artistic output is a highly valuable real-world asset that deserves a secure, sovereign infrastructure of protection.

The Creator Music Ownership Playbook

To build a sustainable career in the modern creative economy, artists must execute a clear, highly disciplined operational playbook.

Playbook Phase	**Core Objective**	**Operational Execution**	**Sovereign Outcome**
1. Intellectual Audit	Establish absolute clarity regarding what you own and control.	Map every single song in your catalog, identifying the exact split of master and publishing rights.	Avoid legal disputes and administrative leaks by having a flawless ledger of your property.
2. Sovereign Distribution	Retain absolute control of your distribution channels and data.	Utilize independent, flat-fee digital distribution platforms rather than signing away rights for advances.	Maintain a direct, unmediated relationship with your streaming revenue and listener demographics.
3. Upstream Structuring	Legally secure your property rights before seeking commercial success.	Establish your own independent publishing administration and register your works with global rights societies.	Ensure that every micro-royalty generated by your sound is tracked, captured, and paid directly to you.
4. Enterprise Cataloging	Transition your focus from single hit songs to building a structured catalog asset.	Bundle your recorded songs and compositions into a single, cohesive intellectual	Build a diversified asset class that can be leveraged, sold, or used as collateral for non-predatory financing.

Playbook Phase	Core Objective	Operational Execution	Sovereign Outcome
		property portfolio.	
5. Legacy Engineering	Align your creative assets with the long-term protection of your family.	Structure your catalog rights within a family trust or digital estate plan to ensure seamless transmission.	Establish a permanent, cash-flowing monument of wealth that supports your descendants for generations.

The Foundation Behind the Creativity

This creative transformation is not an isolated trend; it is part of a massive, global transition in how human value is organized and owned. From the quiet local neighborhoods of Buffalo to the high-visibility fields of elite sports and the late-night rooms of recording studios, the rules of the game remain identical.

Value is created through focused human energy, consistency, and community trust. But if that value is not legally recognized, mathematically measured, and securely structured on a master ledger, it will always be captured by the entities that control the infrastructure.

Sovereignty does not begin with the adoption of a new digital tool or a complex technology; it begins in the quiet terrain of your mind. It is the conscious decision to recognize your value, step out of the passive loop of consumer reaction, and actively build the structured pathways that connect your daily labor directly to a permanent legacy of ownership. Once this internal shift is made, the global systems of the world stop looking like insurmountable obstacles, and begin to reveal themselves as open landscapes of structural opportunity.

The Journey Forward

We have traced the long evolutionary line of ownership from the physical soil of ancient agricultural empires to the industrial machinery of factories, the paper shares of financial corporations, the digital attention-capture of platforms, and the programmable code of decentralized ledgers. We have seen how the Ownership Gap operates as a silent, devastating wedge dividing our society, and we have mapped the practical pathways to bridge that gap in our communities, our small businesses, our sports leagues, and our creative industries.

Now, as our journey nears its destination, we must step back and ask the ultimate, defining question of our era: *Where does the pathway lead next*?

We are standing directly inside a historic inflection point—a permanent, exponential reset of human history. The choices we make today, the literacy we cultivate, and the structures we build will permanently dictate how our children experience tomorrow. The path to sovereignty has been mapped. The light has been turned on. The code is being written. The only question that remains is whether you possess the courage to step onto the pathway, claim your position, and help write the future.

OWNERSHIP PLAYBOOK FOR CREATORS

From Creation → Control

1 KNOW WHAT YOU OWN

- Master
- Publishing
- Brand

Each layer matters.

2 EXPOSURE IS NOT OWNERSHIP

- Attention fades.
- Ownership remains.

Because the game ends.
Ownership does not.

3 STRUCTURE EARLY

Before success — not after.

4 THINK LONG-TERM

A song is a moment.
A catalog is a legacy.

5 EXPLORE NEW MODELS

Not everything—
just enough to understand direction.

6 BUILD BEYOND THE MOMENT

Content creates attention.
Ownership creates permanence.

The Playbook for Creators

Chapter 18
EXPANDING HOW VALUE IS RECOGNIZED

From Impact to Ownership

There are rooms most people will never see. Not boardrooms, not conference halls—rooms where the real work happens. A young person sitting across from you, trying to figure out what comes next. A single parent holding everything together with almost nothing left. A teenager one decision away from a completely different life. There are no cameras in these rooms, no headlines waiting outside the door—just decisions, the kind that quietly shape entire futures.

That is where I spent most of my life working. Not talking about change from a distance, but living inside it every day.

Value That Was Always There

For more than two decades, I worked in Buffalo and Niagara Falls with youth, with single parents, and with young adults transitioning out of foster care. I have seen the work up close: the long hours, the emotional weight, the responsibility of knowing that what you do today can change someone's entire life trajectory.

And one thing was always clear—the value was real. Stronger families, young people finding direction, lives stabilizing. But something else was equally clear: the structure meant to recognize that value was not.

The System Beneath the Mission

Nonprofits operate inside a system that was never fully designed to support long-term value creation. Funding cycles, grants, reporting, and compliance dominate the calendar, creating constant pressure to prove the work instead of expanding it. Leaders spend more time chasing resources than building solutions, forever balancing mission against sustainability, impact against survival.

And here is the hardest truth: some of the most important work happening in our communities is also the most unstable—not because it lacks value, but because that value has never been structured as an asset.

The Lens Must Change

This is where everything shifts. These are not just services—they are assets. Not traditional financial assets, but social assets creating measurable value over time. A youth program that improves graduation rates is an asset. A workforce program that creates stability is an asset. A housing initiative that prevents homelessness is an asset. Because they all do something powerful: they change outcomes. And outcomes, when sustained, create value—economic value, social value, generational value.

The Work I Lived

This became real to me through direct experience. I spent years helping build programs for youth aging out of foster care, guided by one belief that shaped everything I did: if a system removes a child from their family, it has a responsibility not just to protect them, but to prepare them.

When I went back to graduate school, I focused on one question—were young adults leaving these programs truly prepared to live independently? The answer was complicated. Yes, they were learning real life skills: structure, responsibility, how to survive. But survival is not ownership, and that difference matters. Independence is not just paying bills, finding housing, and getting by. It is stability, decision-making, and positioning. Too often, we were teaching them how to live, but not how to own.

The Day I Couldn't Go

May 13, 2013. Barack Obama came to Buffalo, the first Black president in U.S. history. The city was alive, crowds gathering, energy building—a moment people would remember for the rest of their lives.

And I was not there. Not because I did not want to be, but because I was somewhere else.

That same day, there was a ribbon-cutting ceremony for a project completed at 296 Monroe Street: a space built for foster care youth and young parents, a twelve-unit apartment building called the Center for Youth Development. Inside were staff offices, a conference room, a recreational room, a library with computers, and a laundry facility—everything designed to support growth and stability in one place. A one-stop environment.

But to understand what that moment actually meant, we have to step back.

At the time, I was leading the Supervised Independent Living Program. We had scattered-site apartments across Western New York—more than twenty-six in total—and every day, the staff and I were out in the field: driving, responding, managing situations. What I used to call "putting out fires." Not always fires caused by the youth themselves, but by the environments they had been placed in—drugs, gangs, violence, instability. And the truth was, we were spending so much time managing chaos that we had far less time to actually help them build their lives: less time to focus on stability, growth, and self-sufficiency.

At that time, I was not the director yet—I was a coordinator. But I was paying attention. There was another program in the building called Families United. I always liked that name; it meant something. The director and I developed a relationship over time, though I think it really started the day I was walking past their program and a young woman saw me. She needed services. She needed support. And when I looked at her, I realized who she was: my first cousin, someone I loved deeply. I hugged her right there in that moment, and the director saw it. From that point on, our conversations were different—real, grounded, honest.

One day, I came to her frustrated—not with the youth, but with the system around them. I talked about what we were seeing: the barriers, the environments, the constant interruptions that pulled us away from the real work.

And then I did something that changed everything. I stepped back and looked at the system itself. I ran a SWOT analysis—strengths, weaknesses, opportunities, threats—and what became clear was this: the problem was never effort, and it was never care. It was structure. We were trying to stabilize lives inside unstable environments.

So I proposed something different. What if, instead of being scattered, we brought the system together? What if services, staff, and support were all located in one place—a centralized model, a structured environment, a place where youth could not only live, but grow?

That conversation started around 2006. Years later, after navigating a $3.2 million project and learning how to move through the realities of local politics, that idea became real: 296 Monroe Street. And on the day the country was celebrating history, I was standing inside something else entirely—a system we built. Not for recognition, not for attention, but for impact.

Over time, I have seen versions of that model show up in other places, in other organizations, in other systems. I have watched people receive credit for ideas that did not start with them. But that is not what stayed with me.

What stayed with me was this: when systems are built, they require more than ideas—they require alignment. Because as we move into a future where real-world assets are being structured differently, where tokenization begins to redefine how ownership is organized, we are going to need something deeper than innovation. We are going to need cohesion. Not just people working in the same space, but people aligned in purpose.

Because at its best, a system is not just infrastructure. It is a family. And that is where this understanding begins, because the future will not just be built on new technology—it will be built on how well we learn to move together inside the systems we create.

And within those systems, something deeper was taking shape. We were not just delivering services. We were building something more. We were building family.

There was Ms. Sheila—"Mom"—who could pour into you and challenge you in the same breath. Full of wisdom, she spoke life lessons that stayed with you long after the moment passed. Her voice, her laugh, and her smile could light up your day—and gently touch your spirit.

There was Ms. Tracy—"Auntie"—who made one thing clear: education or a vocational path was not a choice; it was a requirement. She showed up wherever she was needed to keep you aligned. Tough when necessary, but always there with whatever it took to see you through to graduation. And she could cook—really cook—the kind of food that didn't just fill you up, but fed your soul.

There was Mr. Paul—"Uncle"—a former military man who taught life and survival skills with precision and discipline. In the moment, it felt demanding—detailed, slow, and unrelenting—but over time, you realize those were the lessons that stayed with you for life. He was strong, but always knew when to be gentle.

There was Mr. Ed—"Grandparent," LCSW—who made it clear that the needs of the youth always came first. Not the system, not the program, not the organization—only the youth. Even when I said we didn't have it in the budget, he never wavered. He advocated for them every time and challenged me to see beyond limitations—to see what was truly needed.

And then there was me—Mr. Teddy, a father figure—standing in the space between leadership and responsibility, knowing every decision carried real consequences. I listened closely to those around me before making the call, balancing safety with understanding and genuine care.

This was not just a team. It was a model—the most dedicated group I have ever worked with. Each of them, I carry with me. Thank you.

Years later, I stayed connected with many of the young people we served. Some became social workers. Some became teachers. Some became professionals. One is now pursuing a doctoral degree. And I asked them something simple: "*What could we have done better?*"

What Was Missing

Two answers came back, clear and consistent.

The first was money. We required savings, and most had between $1,500 and $5,000 by the time they left the program. But almost all of them said the same thing: "*I wish I had saved more.*"

That is when the realization hit me—saving is not enough. Money sitting still does not change outcomes. Ownership does. What if we had taught them how money grows, how assets work, how value compounds over time?

The second answer was deeper: emotional intelligence. Decisions made from fear, from pressure, from uncertainty—and those decisions often shaped their direction more than anything else.

The Missing Layer

Nonprofits are doing powerful work. They build people, stabilize lives, and create opportunity. But there is a missing layer: ownership. Without it, value is created but not sustained. Support is given but not extended. Impact is real but not built to grow.

A Seat at the Table

This is not just about nonprofits. It is about systems. I saw it in conversations around NIL, where decisions were being made without the full voice of those impacted. I saw it working with foundations, often as one of the few voices representing inner-city communities. And I pushed for one thing: youth voice. Because when youth are included, programs improve, policies strengthen, and outcomes increase. No system should be built for people without those people at the table.

A Different Future Is Possible

Now imagine something different. Every young person entering foster care is not just protected, but positioned—with an account, an investment, a structure growing over time. Not just prepared to survive, but entering adulthood with something already building.

This is where systems begin to evolve: value becomes visible, resources become traceable, participation becomes real. This is the shift. From impact to ownership.

From Participation to Positioning

This is where organizations must move—not just doing the work, but understanding the value of the work.

1 — Recognize Impact as Value

Impact is measurable. It is real. It creates long-term outcomes.

2 — Define the Value Clearly

What outcomes are being created? Who benefits? How can it be measured? Clarity turns impact into something that can be structured.

3 — Build Structure Around the Mission

Mission creates purpose. Structure sustains it.

4 — Align Capital With Impact

Not just funding cycles, but long-term alignment between value and support.

5 — Stay Aware of Emerging Systems

Not rushing, but understanding direction.

6 — Stay Grounded in Purpose

Because people, not systems, are the foundation.

The Shift Becomes Personal

At some point, this stops being theory. It becomes real, local, personal. Because the question changes—no longer "*What is happening*?" but "*Who is prepared?*"

The Deeper Truth

Ownership is not built at the top. It is revealed at the ground level—in neighborhoods, in families, in local organizations. Value is everywhere, but ownership is not always aligned with it. And that is the real challenge. Not creation. Connection.

Where the Work Actually Begins

Because the question is no longer theoretical. It becomes practical.

Value → Structure → Participation → Ownership

That is the sequence in the places where people actually live—not in abstraction, but in reality: in nonprofits serving communities, in small businesses building locally, in creators producing value every day, in athletes generating influence, attention, and opportunity. These are not side conversations. They are the system at its most real level.

From Understanding to Strategy

Which means the path forward cannot be passive. It must be intentional. It must be structured. This is where playbooks emerge—not theory, but application: frameworks that help nonprofits connect impact to ownership, small businesses connect operations to assets, creators connect attention to structure, and athletes connect influence to long-term value.

Because value without structure gets extracted. But value with structure can be retained, expanded, and shared.

The Systems Are Already Forming

At the same time, new systems are emerging—quietly, digitally, structurally—changing how value is tracked, how ownership is represented, and how participation is defined. And if communities are not prepared to connect to those systems, they will not disappear. They will simply be positioned within them without understanding.

The Shift in Perspective

Once you understand that communities are not outside the system but foundational to it, the question changes. It is no longer "*Why doesn't the system work*?" It becomes "*What was this system designed to do?*"

Because what often looks like dysfunction from the outside reveals structure when viewed clearly. It reveals pattern. It reveals intention. Not always fair, not always balanced, but consistent. And when you begin to see that, frustration gives way to clarity. Not comfort—clarity. The kind that forces a different kind of thinking, a different kind of strategy.

Because if systems are producing outcomes consistently, then the question is no longer whether they are working. The question is: who are they working for, and how were they designed to function in the first place? Once you see that clearly, you are no longer reacting to the system. You are studying it, positioning within it, building with intention inside it.

Because the truth is, the system was never broken.

THE NONPROFIT OWNERSHIP PLAYBOOK

From Participation ⇒ Positioning

For organizations willing to understand this shift—
there is an opportunity
to move from participation to positioning.

1 RECOGNIZE THAT IMPACT IS VALUE

Impact is measurable long-term value.

- Improved Education
- Reduced System Involvement
- Stronger Community Stability

2 DEFINE AND MEASURE THAT VALUE

Clear outcomes transform impact into value.

- Structure
- Communicate
- Support

3 BUILD STRUCTURE AROUND THE MISSION

Structure sustains mission.

- Clear Program Models
- Defined Outcomes & KPIs
- Financial & Operational Systems

4 ALIGN CAPITAL WITH IMPACT

New models connect capital to outcomes.

- Grants
- Aligned Investment

5 POSITION FOR EMERGING INFRASTRUCTURE

Awareness prepares nonprofits for new systems.

- Transparency
- Efficiency

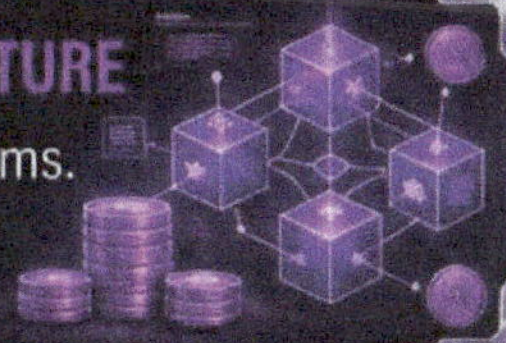

6 STAY GROUNDED IN PURPOSE

Mission is the foundation.

- People
- Community
- Service

The Playbook for Nonprofits

Chapter 19
THE SYSTEM WAS NEVER BROKEN

Whoever Designed This

There were times in my life when I found myself inside systems—working hard, showing up, doing everything I could—and still thinking: whoever designed this did not know what they were doing.

I thought it working with single parents, mothers holding down two jobs, missing sleep, doing everything humanly possible to make life better for themselves and their daughters, and still coming up short—not for lack of effort, but because the system around them never seemed built to let that effort land. I saw it walking side by side with individuals experiencing homelessness, people just seconds away from losing everything, cycling through the same shelters, the same waitlists, the same dead ends, no matter how hard they tried to climb out. I saw it in care coordination, adults stuck in routines, moving through systems they did not fully understand, repeating the same patterns over and over again, year after year, as if the system itself expected them to fail and had simply made peace with it.

And from where I stood, it did not make sense. The effort was there. The people were trying. But the outcomes did not match the effort. So I thought the system was broken.

For a long time, that was the only conclusion available to me. Broken systems, I assumed, produced broken outcomes. If people were working this hard and still falling through the cracks, then somewhere along the way, someone had designed something poorly—carelessly, even. It would take a completely different arena, one that had nothing to do with social work at all, to show me how wrong that assumption was.

What Football Taught Me

But before any of that, I was a defensive coordinator, and football taught me how to see things differently.

Coaching demanded a different kind of attention than most people realize. I was used to breaking down film for hours at a time—studying opponents, watching tendencies, learning strengths, identifying weaknesses, understanding personnel packages. Not just what teams did, but why they did it. Every formation told a story if you were willing to sit with it long enough. Every play call had a history behind it, a pattern that repeated itself once you knew where to look. The film did not lie, even when the players on it were trying to disguise their intentions.

I remember one game clearly. We played a prep school out of North Carolina that had multiple players headed to Division I programs. On paper, they were supposed to dominate. They had the size, the speed, the pedigree—every physical advantage a team could ask for. Anyone looking at the roster on paper would have picked them to win comfortably.

But that night, our defense held their offense to sixty-seven total yards.

Not sixty-seven yards in the first half. Sixty-seven yards for the entire game. After the game, the head coach came up to me, still shaking his head, and offered me a job on the spot. He said, "*No one has ever done what you just did to us.*"

I smiled, I thanked him, and I declined. Because what he did not know was this: I was not surprised.

I Understood His System

I had studied everything—his formations, his tendencies, his personnel groupings, what he liked to run on first down, what he leaned on in short yardage, how he adjusted when he felt pressure building against him. By the time we took the field, I already knew what he was going to try before he called it. So when the game started, it was not chaos to me. It was clear, predictable, structured.

To me, it was an easy game to call—not because I was guessing, not because I got lucky with a few calls, but because I understood the system underneath the system. His players were talented, but talent executes a plan. It does not replace one. And once you understand the plan, the talent stops being unpredictable. It becomes something you can prepare for, position against, and ultimately neutralize.

That is the part most people watching from the stands never see. They see athleticism. They see chaos, speed, collisions. What they do not see is the structure running quietly underneath all of it, dictating almost everything that happens on the surface.

It took years of film, years of losses and near-misses, before that kind of clarity became second nature. Early in my career, I watched games the way most fans do—reacting to what had already happened instead of anticipating what was coming next. It was only after enough repetition, enough hours in a dark room with a remote in one hand and a notepad in the other, that patterns started to reveal themselves faster than the plays could unfold. By the time I was standing across from that North Carolina team, reading a system had become instinct. I was not thinking three steps ahead anymore. I was simply seeing what was already there.

The Realization

And then one day, it hit me the same way it did on the field.

The systems I was working in—foster care, homelessness, care coordination—were not broken. They were running exactly as they were designed to run. Every delay, every barrier, every repeated cycle that I had spent years frustrated by, watching good people fail again and again, was not random, and it was not accidental. It was structured, in the same way that offense out of North Carolina was structured. Someone, somewhere, had built these systems with a set of incentives, priorities, and outcomes in mind, and the results I was seeing were not failures of that design. They were the design working exactly as intended.

That single shift in perspective changed everything about how I approached the work. I stopped asking why people kept ending up in the same place, as if it were some mystery to be solved through more effort or more compassion alone. I started asking what the system had actually been built to produce, and once I asked that question honestly, the answers were no longer confusing. They were consistent. Predictable. The same way an opposing offense becomes predictable once you understand what it is trying to accomplish and why.

Clarity Changes Everything

That is when clarity came. Not frustration, not confusion. Clarity.

Because once you understand that a system is working exactly as designed, you stop asking "*What's wrong*?" and start asking "*What is this system actually built to do*?" That single question changes everything, because now you are not reacting emotionally to outcomes you cannot control. You are seeing structurally, the same way I saw that offense before the ball was ever snapped. You begin studying the incentives instead of just absorbing the results. You begin looking for the pattern instead of getting lost in the individual disappointment of any one case, any one family, any one cycle repeating itself in front of you.

This is the same lens I would eventually bring to ownership, to communities, to the larger systems this book is built around. Because the question was never really about football, and it was never really only about foster care or homelessness either. It was about learning to see structure before reacting to outcomes—a skill that, once learned on a football field, refused to stay there.

Closing Line

I used to think the system was failing people. Now I understand it was doing exactly what it was built to do.

And once you see that clearly, you realize the real work is not fixing the system. It is understanding it, so you can build something different.

Chapter 20
THE COMMUNITY MODEL

Mastercraftsman 110 Degrees Community Framework

Cities as Ecosystems of Ownership

Communities have been misunderstood for a long time. They have been studied, measured, labeled—often described by what they lack instead of what they contain. Researchers arrive with clipboards and surveys, looking for deficiencies to document, and they usually find exactly what they came looking for. But that lens is incomplete, because communities are not problems to solve. They are assets to develop.

Every neighbourhood, every city, every block holds value. Not theoretical value, not potential alone—real value, in the people, in the culture, in the relationships, in the work being done every single day, often without recognition and almost never with the structure needed to sustain it.

The question has never been "Is *value present*?" Value is never actually the question, no matter how many reports suggest otherwise. The question has always been "*Is value being seen, and more importantly, is it being structured?*"

Because when value is not recognized, it cannot be measured. When it cannot be measured, it cannot be supported. And when it is not supported, it is often extracted—quietly, gradually, until the people who created it are the last ones benefiting from it. This is the pattern many communities know well: value is created locally, but ownership lives somewhere else, in a boardroom or a portfolio far removed from the block where the work actually happened. For a long time, this was accepted as normal—until you begin to see clearly.

Before I ever understood systems, before I understood ownership, I had already seen the problem. I just did not have the language for it yet.

When I was younger, everybody in my neighborhood thought they could rap. It was just part of the culture—voices trying to be heard, stories trying to be told, kids turning porches and street corners into stages because no one had built them a real one. That was not me. I was not the rapper. I was the beatboxer, the one creating rhythm so the words could land. Even then, I understood something I could not explain yet: every voice needs structure to be heard.

But there was one time, just once, when I stepped into it myself. My twelve-year-old self said something that stayed with me:

"Yo *homeboy, I had a dream that the whole world became a team. They didn't care about the races, the color of your skin or your faces. All they really wanted was one damn thing—and that was for the whole world to become a team.*"

At the time, it was just a rhyme, something that felt right, something I said and moved past without giving it much more thought. But I did not realize I was echoing something much deeper.

Years before that moment, Martin Luther King Jr. stood in front of a nation and shared a vision of a world where people would be judged by their character rather than the color of their skin—a dream of unity, of shared humanity, of a world that moves forward together. He said he might not get there with the rest of us, but he believed we would. Back then, I did not understand the full weight of that dream. But something in me recognized it anyway, the same way a kid recognizes a rhythm before he can explain why it works.

As I grew older, one thing never changed: I never found it easy to leave people behind. Some would call that a weakness, say you have to separate to grow, to move ahead, to distance yourself from where you came from in order to become something more. But that never sat right with me, because something deeper kept asking: if the system requires you to leave people behind to succeed, what kind of system is it?

That question stayed with me through life, through experience, through observation, until I began to understand something clearly. The issue was never a lack of talent. Never a lack of effort. Never a lack of value. The issue was structure. People were taught how to participate, but not how to own.

And that is where the Community Model begins.

From Value Creation → Ownership Participation → Legacy Impact

1 — Recognize Community as an Asset (Awareness)

Communities are not empty. They are not behind. They are not broken. They are unstructured value. Every community already holds talent, creativity, work ethic, cultural identity, and social capital. The shift begins when we stop asking "*What is missing?*" and start asking "*What already exists, and how do we structure it?*"

2 — Make Value Visible (Clarity)

You cannot build ownership around what is not seen. So the next step is visibility: what value is actually being created? Youth development outcomes, small business growth, community engagement, creative production, local economic activity. When value becomes visible, it can be measured, it can be supported, and it can be structured.

3 — Build Ownership Structures Around Value (Structure)

This is where everything changes. Value alone is not enough. Structure determines who owns, who benefits, who participates, and who controls. Without structure, value created in a community can be captured outside of it. With structure, value stays aligned with the people creating it.

This is where tools begin to matter—not as hype, but as infrastructure: tokenization, digital asset frameworks, new financial models. Not to replace systems, but to make ownership more visible, more accessible, and more aligned with the people generating it in the first place.

4 — Enable Ownership Participation (Activation)

Participation must evolve—from working, consuming, and renting, to owning, earning, and sharing in outcomes. This is how communities move from activity to stake, where people are no longer just involved. They are positioned, with a real claim to what they helped build.

5 — Keep Value Circulating Locally (Alignment)

This is where sustainability begins. When ownership is structured correctly, value does not leave. It circulates. Communities benefit from what they build, and individuals grow with what they contribute to. This creates stability, resilience, and long-term growth—not temporary progress, but self-sustaining ecosystems that do not need to be rebuilt from scratch every generation.

6 — Strengthen Systems and Governance (Integrity)

As value grows, structure must mature, because without governance even strong systems can break. This stage ensures transparency, accountability, and community-led governance: clear rules, shared responsibility, ethical stewardship. Not control, but aligned leadership—where systems are not only built, but protected, maintained, and trusted. Ownership without integrity creates instability. Ownership with governance creates confidence and continuity.

7 — Create Legacy and Intergenerational Impact (Completion)

This is the highest level of the model, where value is no longer short-term but transferred across time. Legacy is not what is left behind. It is what is structured to continue. This includes knowledge transfer, long-term ownership structures, wealth preservation, and future generation participation—where communities are no longer reacting, but building forward, and where children do not start from zero, but from position. This is how value becomes enduring, transferable, and expanding—not just for today, but for generations.

When value is recognized, structured, aligned, and sustained, communities move from participation to ownership to legacy.

What This Means Moving Forward

That dream I spoke as a kid, about the world becoming a team, was never just about unity. It was about alignment. Because a real team shares responsibility, opportunity, and outcomes—not just effort. And for the first time, we are entering a moment where that kind of alignment can actually be built. Not just imagined. Built.

Dr. King gave the world a vision, a direction, a truth. What he did not have were the tools to structure that vision into systems that could outlast a single generation. We do. And that is why this moment matters—not because the dream has changed, but because for the first time, the infrastructure to build it finally exists.

Because now the question is no longer "*Can communities rise?*" They already are, and they have been all along, often without the recognition or the structure to show for it. The real question is: "*Will they own what they build?*"

In the end, this was never just about helping communities participate. It was about making sure they are finally structured to own—so that the next version of that dream is not just spoken by a twelve-year-old on a porch, but built into something that lasts.

HE MASTER CRAFTSMAN 110° COMMUNITY MODEL FRAMEWORK
From Value Creation → Ownership Participation
1 RECOGNIZE COMMUNITY AS AN ASSET
Talent • Creativity • Work Ethic • Culture • Social Capital
From "What's Missing?" To "What Already Exists?"
2 MAKE VALUE VISIBLE
Youth Outcomes • Small Business Growth
Community Impact • Local Economy
MAKE VALUE VISIBLE
3 BUILD OWNERSHIP AROUND VALUE
Who Owns • Who Benefits • Who Controls
Tokenization • Asset Models
Tokenized Assets
Asset Models
Ownership Participation
Fractional Ownership
4 ENABLE OWNERSHIP PARTICIPATION
From Working
Consuming
Renting
Community = Unstructured Value
To Owning
Earning
Sharing
5 KEEP VALUE CIRCULATING LOCALLY
Local Reinvestment • Shared Outcomes
Sustainable Growth
6 STRENGTHEN SYSTEMS & GOVERNANCE
Transparency • Accountability • Community
Governance • Clear Rules • Ethical Stewardship
7 CREATE LEGACY & INTERGENERATIONAL IMPACT
Legacy Building • Knowledge Transfer
Long-Term Ownership • Future Generations
When value is recognized, structured, and aligned—
Communities move from participation... to ownership.
The Game Plan for Community Ownership
EMPIRE NEXUS
MAKING THE UNSEEN UNDERSTOOD

Chapter 21
BUFFALO, REALITY AND RESPONSIBILITY

Seeing Clearly Before We Build

I have seen what happens when people are expected to move before they understand. They hesitate. They react. They fall back on what feels familiar. Not because they lack intelligence, but because clarity was never established. And when clarity is missing, even the strongest people can appear uncertain, even the most capable communities can feel stuck, cycling through the same reactions to the same pressures without ever getting underneath what is actually happening to them.

This is not new. Buffalo has lived it.

Buffalo is not a theory. It is a reality—a city built on effort, on discipline, on people who understand what it means to carry weight over time. Generations who showed up day after day, not because it was easy, but because it was necessary. There is a particular kind of toughness that comes from winters that do not ask permission and industries that do not wait for you to catch up. Buffalo has that toughness built into its bones.

There was a time when Buffalo stood at the center of movement: the Erie Canal, steel, manufacturing, trade. This city did not just participate in progress—it helped define it. Energy moved through Buffalo. Goods moved through Buffalo. Opportunity moved through Buffalo. This was not a place on the sidelines watching the economy happen elsewhere. This was the system.

And then the system changed—quietly at first, then all at once. When industry left, it did not just remove jobs. It removed structure. It removed pathways. It removed the visible connection between effort and long-term opportunity that an entire generation had built their lives around. What once felt stable, predictable, understood, no longer made sense. And when structure disappears, people are left to navigate without a map, often blamed for struggling to find their footing in terrain that shifted without warning.

But the people remained. They stayed, they adapted, they adjusted, they survived. They found ways to provide, ways to protect, ways to keep going, even when the system around them was no longer moving with them. That matters. That is strength. That is identity. That is Buffalo.

But survival is not ownership. Survival is reactive; ownership is structured. Survival responds to what is happening; ownership positions for what is coming. And that difference is everything, because one keeps you standing in place while the other moves you forward.

A Different Kind of Shift

Now another shift is here, but this one is different, because this is not just a shift in industry. This is a shift in function. In the past, machines replaced physical labor—factories closed, jobs disappeared, communities adjusted, and eventually new forms of work emerged to replace what was lost. There was pain in that transition, but there was also a landing place on the other side of it.

This time is different. We are not just replacing labor. We are replacing the need for labor itself. Artificial intelligence, automation, robotics—systems that do not just assist, they operate. Faster, more efficient, more scalable than anything that came before. Not competing with effort, but redefining whether effort is even necessary in the first place.

Buffalo has seen disruption before. When jobs leave, communities feel it first, long before economists write the reports explaining why. When systems change, people are expected to adjust after the fact, quietly, without much support, and mostly on their own. But this time the scale is different. This is not one industry. This is not one city. This is global. This is systemic. This is foundational.

And now the question becomes unavoidable: if human labor is no longer the primary way people participate in the economy, then how will people participate at all? If jobs become limited or unnecessary for entire categories of work, what becomes the pathway to income, stability, growth, and ownership?

Because if participation disappears and no new structure replaces it, people are not just displaced. They are disconnected—disconnected from value, disconnected from systems, disconnected from opportunity itself. That is not a temporary issue that resolves itself with enough patience or enough retraining programs. That is structural.

What Buffalo Teaches Us

And Buffalo teaches us something critical: people will adapt. They always do. That has never been in question. But adaptation without structure leads to survival, not ownership. And survival, over time, becomes limitation, a ceiling that gets lower with every generation that inherits the same unstructured position instead of something better.

So the real question is not whether people will adjust—they will, they always have, out of necessity if nothing else. The real question is: what are they adjusting into?

Because if the next system is built without pathways to ownership, then history repeats. People will contribute without accumulating. Participate without positioning. Work inside systems without ever benefiting from how those systems grow, watching the value they helped create end up somewhere else entirely.

And that is the moment we are standing in now. Not fear. Responsibility. Because for the first time, we are seeing the shift while it is happening—not after, not years later, not through a retrospective report explaining what should have been done differently. Now, while there is still time to choose differently.

And that changes everything, because now we have a choice: to build differently, to structure participation intentionally, to connect value to ownership before the system fully settles into its next shape.

Because if labor is no longer enough, then ownership must become the foundation—ownership of assets, ownership of systems, ownership of value creation itself. Not just working inside systems, but being positioned within them in a way that survives whatever comes next.

This is where the future begins to take shape. Not in jobs alone, but in ownership. Not in effort alone, but in structure. Not in participation alone, but in who benefits over time, long after the initial effort has been forgotten.

The Questions We Cannot Avoid

Buffalo has already lived the lesson. We have seen what happens when systems change and people are left to figure it out later, on their own, without the tools they needed the first time around. The pattern is not hidden. The lesson is not new. The only question now is: will we see it clearly before it happens again?

Because before anything can be built, it must first be understood. And before people can own, they must first see.

And speaking from where I stand, as someone shaped by this city, as someone who understands what it means to earn everything, I can say this clearly: we were never given anything. We worked for it. We earned it. We carried it. That mindset is not just cultural. It is foundational to how I see everything in this book.

So when I hear conversations about abundance, when I hear ideas like universal basic income, I do not immediately hear opportunity. I hear questions—real questions about structure, about sustainability, about control, about ownership. Because a check without a structure underneath it is still just survival, dressed up differently.

Because if the future is moving beyond labor, then what replaces it must be understood. Not assumed. Not accepted blindly because it sounds generous. Understood, the same way Buffalo eventually had to understand what replaced the factories, except this time we have the chance to ask the questions before the transition happens instead of after.

And those are the questions we cannot avoid. Those are the questions we must answer. Because what comes next will not just shape systems. It will shape lives, in the same neighborhoods, on the same blocks, in the same city that has already had to learn this lesson once before.

And that is where we go next.

Chapter 22
THE NEW CITIES OF POWER

Who Builds the System... Owns the Future

Musk and the Return of the Company Town

It does not always begin with an announcement. Sometimes it begins with land—quiet purchases, wide spaces, infrastructure placed far from noise. Then, over time, something begins to take shape. Not just buildings, not just roads. A system.

Musk did not set out to build a city in the traditional sense. He built what he needed: a launch site, a testing ground, a place where ideas could move faster than resistance. And then something shifted. Because once you control the land, and the company, and the workforce, you are no longer just building a business. You are shaping an environment.

Starbase, Texas became real—a place where work happens, people live, infrastructure is controlled, and the entire system is aligned to one central force. Snailbrook followed the same pattern: housing for employees, schools for families, life organized around the system.

This is not new. It is familiar. But it has returned in a more advanced form.

The Pattern Beneath the Surface

Look beyond the names. This is not about one person. This is about a pattern repeating itself: the company town, but now powered by technology, data, automation, and capital at scale.

The old company town of the twentieth century controlled where you lived and where you shopped. It rarely controlled how you communicated, what you saw, or how your work itself was measured in real time. This new version does. These are not just environments. They are integrated systems, wired together in ways the mill towns and mining towns of the past never had the technology to achieve. And the people building them are not just employers. They are system designers, shaping every layer of daily life at once—housing, employment, information, and increasingly, the infrastructure that decides what opportunity even looks like inside their walls.

Telosa and the Idea of Redesign

Not every builder is moving quietly. Some are trying to redesign the system openly. Marc Lore introduced Telosa, a city imagined differently, where land is not concentrated, where value is structured intentionally, where ownership could be shared. Whether it succeeds or not is not the point. The idea matters, because it shows that even at the highest levels of wealth and influence, people are questioning the structure of ownership itself.

Silicon Valley and Controlled Ecosystems

Others are building quietly—expanding campuses, adding housing, adding retail, adding transportation—until eventually an entire ecosystem forms. Where people work, live, spend, and move inside a system they did not design and rarely get to question.

The Buffalo Moment

And this is where it became real for me. Not in Texas. Not in Silicon Valley. But in Buffalo. Because I have lived inside systems that were already designed, just not explained.

I have walked neighborhoods on the East Side and the West Side where talent was everywhere, but ownership was nowhere to be found. I have sat in rooms with young people who had everything it takes: discipline, creativity, resilience. But the system around them was not built for them to own anything. Only to participate.

I remember being in a community center, watching a group of kids run full speed across a gym floor that had seen better days—worn hardwood, flickering lights, a scoreboard that had not worked in years. The energy in that room was real. The potential was real. But something else was real too: the structure around them did not match the value inside them.

That is when the question hit me differently. Not just "*What do these kids need?*" But "*What system are they inside of?*" Because if the system does not change, the outcome rarely does.

Buffalo taught me something no book could: systems do not have to be announced to shape your life. They are already operating, in housing, in education, in access, in economics. And if you are not part of how the system is structured, you are part of how it is experienced.

That is when everything began to shift for me. Because I started to see that the conversation is not just about opportunity. It is about ownership of the system itself.

The Shift That Is Actually Happening

Now look back at what is happening across the country. Billionaires building cities. Companies building ecosystems. New models being tested. This is not random. This is the next layer of control.

The New Structure

First comes land. Then comes capital. Then comes infrastructure. Then comes people. And eventually, governance follows. Because once enough of life is contained within a system, the system begins to define reality.

The Missing Layer

There is one layer not yet fully visible. But it is coming: digital ownership. Because once systems are built, the next question becomes who owns them, who participates in them, and who benefits from them. Right now, ownership is still concentrated. Participation is wide. But the gap between the two is where the future is being decided.

Two Directions From Here

This is where the path divides.

Path One — Controlled Systems

People live inside corporate ecosystems, centralized infrastructure, pre-designed environments. Participation exists, but ownership does not expand with it.

Path Two — Structured Ownership

Systems are built where people can participate, people can own, and people can benefit. Value created inside the system is shared, structured, visible.

This is where tokenization enters—not as hype, but as structure. A way to make ownership divisible, traceable, and transferable. Not just for music. Not just for assets. But eventually, for entire systems.

The Real Question

The question is not whether cities will be built. They already are. The question is not whether systems will be controlled. They already are. The real question is: when new systems are built, will communities like Buffalo be included in ownership, or only in participation?

Because I have seen what happens when people are locked out of ownership. And I have seen what happens when awareness begins to change that.

Closing Line

This is not about cities. It is about systems, and who they are built for. And the future will not be defined by who lives inside these systems, but by who owns a piece of them.

Because while new cities are being imagined, designed with precision, built with advanced technology, and positioned for the few who can access them, something else is being offered to everyone else. Not ownership. Not control. A promise. A promise of abundance—a world where technology provides, where systems distribute, where life is sustained without the need to own what produces it.

For many, that promise will feel like progress. But beneath it, a deeper question remains. Who controls the systems that provide that abundance? Who decides how it is distributed? And what happens when participation no longer requires ownership, but dependence?

Because the future may not divide between those who have and those who don't, but between those who own the systems and those who live within them.

And that is where this conversation leads next.

Chapter 23
THE PROMISE OF ABUNDANCE

Technology, Power, and the Question We Must Ask

A Powerful Idea

What if the future does not take from you, but simply removes your need to own anything at all? This is not a distant possibility. This is a direction already forming.

There are voices today speaking about a future of abundance. A future where artificial intelligence and robotics eliminate scarcity. A future where work is no longer necessary. A future where income is no longer a concern, because there will be what some are calling universal basic income.

It is a powerful idea. A hopeful idea. A vision of a world where survival is no longer the primary concern, and where humanity is free to live, create, and evolve without constraint.

But when I hear it, I do not react with fear. I do not reject it. I do not blindly accept it either. I pause. And I ask questions.

Because history has shown us something clearly: every major technological shift carries two things at the same time—opportunity and consequence. The future will be no different.

So the question is not simply, what is possible? The question is: who controls it? Who benefits from it? And how will it shape everyday life? Because systems are not shaped by possibility alone. They are shaped by incentives.

The Deeper Question

And that leads to a deeper question, one that cannot be ignored: if we are moving toward ideas like universal basic income, what would truly incentivize someone to give their riches away to people all over the world?

That question stayed with me. Not as doubt, but as something unresolved—something that required more than a surface-level answer. So I sat with it. I slowed down. And I went deeper.

With Luke, for seven days, we talked it through. We challenged it. We questioned every angle. And only after those seven days with Luke, verse four, did it begin to make sense.

Because once you move past the surface, another layer reveals itself. If we already have the resources today to reduce suffering, why does it still exist? If systems are already generating massive wealth, why is access still limited? And if the most powerful systems in the world today are still driven by growth, control, and shareholder value, what fundamentally changes in the future that would suddenly shift those incentives?

These are not questions of skepticism. They are questions of awareness. Anyone who has spent time inside large institutions knows how difficult it is to change an incentive structure once it is already producing results for the people at the top of it. Wealth does not typically redistribute itself simply because the technology to do so becomes available. It redistributes when the structure requires it to, and right now, very little of what is being built requires it.

Systems and Incentives

Because as technology evolves, as artificial intelligence expands, as automation scales, we are not just building tools. We are building systems. And every system carries structure. Every structure carries power. And every form of power raises a question: who controls it, and who participates in it?

Centralized systems offer efficiency, coordination, and scale. But they also introduce risk: control, surveillance, dependency. Decentralized systems offer participation, transparency, and access. But they require something different: responsibility, understanding, engagement.

And this is where the conversation must rise. Because this is not about choosing sides. It is about choosing awareness.

After reflection, after questioning, after sitting with what is being built, something became clear. The future is not guaranteed to be one thing. It is not automatically abundant. It is not automatically controlled. It will be shaped by human choices, human values, human awareness.

Technology can create abundance. But it can also concentrate power. It can free people, or it can condition them. And the difference will not be determined by the technology itself, but by how humanity chooses to build around it.

What Abundance Actually Requires

That is when the realization deepened. The conversation is not really about whether abundance is possible. It is. The real question is: what kind of systems will we build to support it? Systems rooted in trust, or systems designed around control?

Because even in a world of abundance, ownership still matters. Participation still matters. Awareness still matters. Abundance without ownership can become dependency. Abundance without understanding can become control. Abundance without participation can become illusion. But abundance with awareness changes everything.

Because when people understand, fear turns into curiosity. Confusion turns into clarity. Distance turns into participation. And once that shift happens, a new question emerges. Not "*What is happening*?" But "*What can I do*?" And that question is where everything begins to change, because that is the moment where awareness becomes action.

From Awareness to Action

And this is where everything becomes real, because once the question shifts from "*What is happening*?" to "*What can I do*?", something deeper begins to take place. Awareness expands, and with it, responsibility. Because it is one thing to see the system. It is another thing entirely to understand how to move within it.

This is where many people stop. Not because they are incapable, but because no one ever showed them the path. They are told to work, told to adapt, told to "keep going."

But rarely are they shown how to build within change. And without that, awareness can become overwhelming. Clarity can turn into hesitation. Understanding can stall before action ever begins.

That is the gap. Not just a gap in income, not just a gap in access, but a gap between seeing and building, between understanding and applying, between awareness and ownership. Because if the future is shifting, then people do not just need information. They need structure—a way to move, a way to position, a way to participate that is clear enough to follow and strong enough to sustain.

And this is where the conversation must evolve again: from ideas to execution, from awareness to design, from possibility to pathway. Because if ownership is going to expand, it will not happen by accident. It must be taught, structured, and built intentionally—not just for institutions, not just for those already positioned, but for communities, small businesses, athletes, creators, and everyday people who are already creating value without always being connected to it.

Because the truth is, the value already exists. The effort already exists. The potential already exists. What has been missing is the structure to connect it.

And once that becomes clear, something powerful happens. The conversation stops being about the future and starts becoming about what we build next. Because awareness has already done its job. You can see the system. You can feel the shift. You understand what is at stake.

Now the only question that remains is: how do we build something people can actually use? Not theory. Not concepts. Not conversations that never leave the room.

But something real, something structured, something that can move people from participation to ownership.

And that is where we go next.

Chapter 24
THE COMMUNITY OWNERSHIP GAME PLAN

From Awareness to Action: The Ownership Framework in Motion

The Path Forward

Everything up to this point has been about seeing clearly—seeing the systems, seeing the patterns, seeing the gap between value and ownership that runs through Buffalo, through nonprofits, through cities being built by billionaires, through the very idea of abundance itself. Seeing is necessary. But seeing alone does not change a single outcome for a single family. At some point, the conversation has to turn from what we understand to what we actually do with it.

The path forward is clear. Not complicated, not theoretical. Clear.

Because once awareness is established, the question is no longer what is happening. The question becomes: how do we build something people can actually follow?

And the answer is this: not more information, not more complexity. Structure. A clear, repeatable pathway that moves individuals and communities from awareness to ownership.

This is that pathway.

The Community Ownership Framework

1. Awareness — See Clearly First

Everything starts here. Before strategy, before systems, before action, people must understand what is happening.

So we meet people where they already are: churches, schools, community centers, neighborhood spaces. Not a rented conference room downtown that requires two bus transfers to reach, but the places people already trust enough to walk into on a Sunday or a Tuesday afternoon.

We do not overwhelm. We do not confuse. We do not sell. We explain—clearly, simply, honestly. No hype, no fear.

Because once people understand, they begin to see. And once they see, they cannot unsee.

2. Understanding — Make It Real

Big ideas do not need to be complicated. Money, technology, ownership—these are not complex; they have just been explained poorly, usually by people who never had to sit across from someone deciding between rent and groceries.

So we change how they are taught. We use simple language, real-life examples, everyday situations. We connect concepts to lived experience instead of abstractions borrowed from a finance textbook.

Because when people understand something, they are not intimidated by it. They can use it.

3. Trust — Deliver Through Familiar Voices

Information alone does not move people. Trust does.

So the message must travel through people who are already trusted: coaches, teachers, church leaders, mentors, community leaders. People who already show up. People who already care, long before there was any framework to hand them.

Because when trust is present, people listen differently. They open. They engage. They consider. And that changes everything—not because the information itself is new, but because the messenger has already earned the right to be heard.

4. Action — Make It Easy to Start

Awareness without action creates frustration. Because once people understand, they will ask, "*What can I do?*" And we must have an answer ready, not a vague promise to follow up later.

So we create simple, clear entry points: workshops, ownership-based learning sessions, community investment ideas, small business participation pathways, collaborative local projects.

We remove friction. We simplify the starting point. We make it easy to understand, easy to join, easy to begin.

Because action builds confidence. And confidence builds momentum.

5. Ownership — Build It Together

We do not build for people. We build with people.

Ownership is not just access; it is participation in creation. Everyone has a role. Everyone has a voice. Everyone has the ability to contribute, regardless of how much capital they started with.

This is how systems last. This is how communities grow. This is how value stays connected to the people creating it, instead of drifting toward whoever happened to be positioned to capture it.

Because when people are part of building, they do not just participate. They own.

What This Looks Like in Practice

Picture how this actually moves through a neighborhood. It starts small: a conversation after Sunday service, a workshop announced on a church bulletin board, a coach pulling a few players aside after practice to talk about something beyond the game. No stage, no slide deck, no pressure to commit to anything on the spot.

From there, the same handful of people who showed up start bringing others. A parent tells another parent. A teacher mentions it to a colleague. The circle widens not because of marketing, but because the message is carried by people the community already trusts, and because what they are hearing finally makes sense in a way it never did before.

Then comes the first real step: a small workshop, a modest investment idea, a chance to participate in something local at a scale that does not feel intimidating. Nobody is asked to bet everything on an unfamiliar concept. They are asked to take one clear, understandable step, and then another.

That is how awareness becomes ownership—not in a single dramatic leap, but through a sequence that any person, in any neighborhood, can actually follow from wherever they happen to be standing.

What Makes This Different

This is not a program. This is not a one-time initiative. This is not a short-term solution that disappears the moment the grant cycle ends.

This is a model—a repeatable structure that can be applied across communities, schools, nonprofits, athletic programs, and small business ecosystems. Anywhere value is being created but ownership is not clearly connected.

Because the truth is, the value is already there. The effort is already there. The people are already there. What has been missing is the structure to connect it.

The Shift This Creates

When this model is applied, something begins to change, and it rarely announces itself with a single dramatic moment. It shows up gradually, in the tone of conversations, in who starts asking different questions.

People stop feeling disconnected. They start seeing opportunity. Communities stop reacting. They start building. Participation stops being temporary. It becomes intentional. And ownership—something that once felt reserved for people far removed from the neighborhood—becomes possible for the people actually standing in it.

This Is Where It All Comes Together

We began this journey with awareness—understanding systems, recognizing change, seeing what was always there. Now we arrive at something different: not just understanding, but application.

Because the future will not be shaped by those who simply observe change, but by those who understand it and choose to build within it. And that is the opportunity in front of us now. Not to wait, not to react, but to build—clearly, intentionally, together.

Because ownership is not something reserved for a few. It is something that can be understood, something that can be structured, something that can be built when people are given a clear path to follow.

And now, that path exists.

THE COMMUNITY OWNERSHIP GAME PLAN

From Awareness to Action (Simple and Clear)

The path forward is clear.
Not confusing.
Not complicated.
Clear.

1 START WITH AWARENESS

Everything starts here.

- Before plans.
- Before systems.
- Before action.

People need to understand what's going on
So we create simple spaces to learn.

Places people already go: • Churches • Schools • Community centers

2 TEACH IN A WAY PEOPLE UNDERSTAND

Big ideas don't have to be hard.

- Money
- Technology
- Ownership

Use: Simple words • Real-life examples • Everyday conversations

3 USE PEOPLE THEY ALREADY TRUST

People don't just listen to information.
They listen to people they trust.

People who already show up. • People who already care.

4 GIVE PEOPLE SOMETHING THEY CAN DO

Learning is not enough.
People need a way to act.

We make it easy to start. Easy to

5 BUILD IT TOGETHER

We don't build.

We don't build this for people.
We build it with people.

This is how it lasts. This is how it **grows.**

6 THE GAME PLAN FOR COMMUNITY OWNERSHIP

Tokenization is a tool that can help communities turn value into ownership.

instead of value being hidden or temporary—it becomes organized, trackable.

The Game Plan for Community Ownership

EMPIRE NEXUS
MAKING THE UNSEEN – UNDERSTOOD

Chapter 25
THE COST OF INACTION AND MISALIGNMENT

The Question We Cannot Soften

There is a part of this conversation that cannot be softened. Because understanding the future is not enough. We also have to face this: what happens if we don't move with it?

My friend, who is more like a brother, says it simply: "*We walking with faith, not fear.*"

Because fear keeps people still. Fear delays decisions. Fear keeps people locked inside systems they do not understand. But this is not about fear. This is about clarity.

Because whether we act or not, the system is still moving. Technology is advancing. Ownership is restructuring. Value is reorganizing itself in real time. And if we are not aligned with that shift, there is a cost. Not theoretical. Not future. Structural.

The Cost of Inaction

Inaction does not mean nothing happens. It means everything happens without you.

Systems will continue to produce value, but you will not be connected to it. Opportunities will expand, but you will not be positioned for them. Access will exist, but you will not understand how to use it. Think of the nonprofit still chasing the same grant cycle it chased a decade ago, or the small business owner still trading hours for dollars while entire categories of value shift onto platforms they have never looked at. Nothing dramatic happens to them. That is exactly the problem.

And slowly, without announcement, you become external to the system. Not removed. Not excluded by force. But disconnected by design. Still working. Still contributing. But no longer participating in how value grows.

That is the shift. Not visible at first. But undeniable over time. By the time it is visible, it usually shows up as a feeling more than a fact—a sense that everyone else seems to be moving faster, building more, connecting to something you cannot quite name yet.

The Cost of Misalignment

Misalignment is more dangerous than inaction, because misalignment feels like progress. You are moving. You are working. You are doing everything you were taught to do. But the system has changed, and you have not. So your effort no longer connects to the outcomes you expect.

You work, but do not build. You earn, but do not accumulate. You participate, but do not own. And over time, the gap widens. Not because of effort. Because of structure. Because the system is no longer designed to reward what you are giving it in the way it once did.

And if that continues, you do not fall behind suddenly. You fall behind systematically, one missed alignment at a time, until the distance is too wide to close with effort alone.

The Part People Avoid

There is a side of this conversation most people do not want to look at, because it is uncomfortable. But it is real.

Every system has two sides: what it creates, and what it extracts. What it expands, and what it concentrates. What it makes visible, and what it hides.

Ownership is not just opportunity. It is power. And power, when unexamined, concentrates. If awareness does not expand, ownership will not expand. It will consolidate—into fewer hands, fewer systems, fewer points of control. That is not a prediction about some distant future. It is simply what unexamined power has always done, in every era, under every kind of system.

And in that world, people will still live, still work, still participate. But they will not be positioned. They will exist inside systems without influence over them.

This Is Where It Turns

This is not about fear. This is about truth.

Because if we do not build awareness, someone else will build the system. If we do not structure participation, someone else will define it. If we do not expand ownership, it will not remain neutral. It will concentrate.

That is not opinion. That is pattern. And that pattern has already happened before, in Buffalo and in cities like it, in industries that rose and fell without ever including the people who built them from the ground up.

The Real Consequence

The risk is not that people fail. The risk is that people continue without realizing the system has changed—working inside a model that no longer connects them to value, believing they are progressing while the structure moves past them.

And by the time it becomes clear, the system is already established. And much harder to change.

What This Moment Requires

Not panic. Not reaction. Alignment. Clarity. Courage.

The courage to see what is actually happening. The courage to question what feels normal. The courage to move before everything is fully explained. Because this moment does not wait, and the system will not pause for understanding. It will continue to build, with or without awareness, with or without participation.

The only question is: will you be aligned with it, or positioned outside of it? Because what we choose not to see will still shape what we experience. And what we choose not to build will still be built. Just not by us.

And that is the cost.

Where This Takes Us Next

If we are going to move forward, we cannot stay at the surface. We have to go deeper—beyond what is visible, beyond what is comfortable, into the structures, into the systems, into the parts most people do not fully see, or choose not to question.

Because what happens next will not be determined by what is obvious, but by what is understood, or ignored. And that is where the real divide begins. Not between those who have more, but between those who see clearly and those who do not.

Because when systems evolve without awareness, they do not just change outcomes. They lock them in. And once that happens, the ability to adjust, to participate, to reposition becomes smaller with each passing moment.

And that leads us to the question that cannot be avoided. Not what happens if we succeed, but what happens if we don't.

Chapter 26
WHAT HAPPENS IF WE DON'T

Automation, Control, and the Future Without Awareness

Work Disappears

Everything discussed so far in this book—the Community Model, the Ownership Framework, the cost of inaction and misalignment—has been building toward this chapter, because none of it matters if we do not first understand what happens when none of it is applied. This is the version of the future that forms by default, not by design, when awareness never expands past the surface of daily life.

This is not projection. This is direction. This is what happens when awareness does not expand.

Work disappears. Not suddenly, not dramatically, but systematically. Automation replaces labor. Robotics replace repetition. Artificial intelligence replaces decision-making—not partially, but structurally.

Entire industries shift out of human necessity: transportation, manufacturing, logistics, customer service, administration. Even areas once believed to be protected—education, healthcare—are not immune. What once required people no longer does. And this is not a distant hypothetical confined to factory floors; it is the same pattern already visible in the call centers, the warehouses, and the back offices quietly shrinking their headcount every year.

And without awareness, people are not prepared.

Income Without Earning

If labor is no longer required, income must be reassigned—not through effort, but through systems: programs, distribution models, universal structures. Access to money becomes detached from production. And when income is no longer earned, it is no longer controlled. It is received, on terms designed elsewhere, by people who did not have a seat at the table when those terms were written.

When income is distributed this way, dependency is not a side effect. It is the structure. People rely on systems for stability, for access, for survival. And reliance without understanding removes leverage. There is nothing to negotiate, nothing to reposition, nothing to build outside of. Only something to receive—and something that can be adjusted, quietly, by whoever is holding the controls.

Compare that to the difference between savings and ownership discussed earlier in this book. A young person leaving foster care with a few thousand dollars saved had something small, but it was theirs, earned and controlled by their own decisions. A person receiving a distributed income has something larger, perhaps, but it belongs to the system that grants it, not to the person receiving it. That distinction—between what you earn and control versus what you are given and can have taken away—is the entire difference between ownership and dependency.

Access, Surveillance, and Redefined Freedom

If systems distribute value, they also define access—when, how much, under what conditions. Access is not removed. It is managed, adjusted. And without awareness, that management is not recognized. It becomes normal, the same way any inconvenience becomes normal once enough time passes without anyone questioning it.

Automation does not operate in isolation. It observes, tracks, records, and measures continuously—behavior, movement, transactions, engagement—not occasionally but structurally, because systems require data to function. And without awareness, observation is not questioned. It is simply adapted to.

When systems control income, access, and participation, freedom does not disappear. It changes form. Access becomes conditional. Participation becomes behavioral. Opportunity becomes aligned with compliance—not forced, but structured. And without understanding the structure underneath it, movement feels free. But it is limited, the way a leash still allows movement right up until it doesn't.

Participation Without Position

People remain inside systems, but without influence. They receive outcomes without shaping them. They engage without ownership. They are included, but not positioned.

A divide forms—not temporary, but structural—between those who understand systems and those who do not. One group builds, owns, and designs. The other uses, adapts, and depends. And without awareness, that divide does not close. It stabilizes, generation after generation, until it starts to look less like an accident and more like a permanent feature of the landscape—something inherited the way a neighborhood or a last name is inherited, without anyone ever voting on it.

System design, algorithmic control, access rules, distribution structures—these decisions are made without visibility. And without understanding, they are not questioned. They are simply experienced, the same way weather is experienced rather than negotiated with.

The Shift Has No Announcement

There is no moment where everything stops. No clear line between before and after. Only progression. Jobs fade. Systems expand. Dependency increases. Until one realization emerges quietly: people are no longer participating in systems. They are being managed by them.

Without awareness, there is no preparation, no positioning, no leverage, no influence. Only response. And response is always behind structure.

This is the outcome: people depending on systems they do not understand, receiving income they do not control, living within structures they did not shape, observed without questioning why, operating within limits they did not define. And over time, this will not feel restrictive. It will feel normal.

The Final Truth

If awareness is ignored, automation does not free people. It places them inside systems they cannot influence.

Everything leads here. Not information, not theory. Position. Because from this point forward, this is no longer a concept. It is a decision, a direction, a path.

Because when everything becomes clear, the conversation simplifies. Not easier, but sharper. You stop asking what is happening and start asking what it means, how it connects, how it holds together. Because beneath every system, no matter how complex it appears, there is always a structure, a pattern, an equation—something that explains how it works and why it leads where it leads.

And once I saw that, I could not look at any of this the same way again. Because this was not random. It was not disconnected. It was not out of control. It was following something—something I recognized, something I trusted long before I ever had the language for systems.

That trust in structure, in the idea that something logical sits underneath even the most chaotic-looking situation, did not begin with any of this. It began much earlier, with something far simpler than automation or income distribution. It began with numbers.

Chapter 27
THE EQUATION I COULDN'T IGNORE

The Big Three and the System That Solves Them

The First Thing I Ever Trusted

I have always been a numbers person. Before I understood systems, before I understood ownership, I understood math.

I remember being in eighth grade, traveling to math competitions across Western New York. Different schools, different rooms, different problems to solve. But I always looked forward to one place in particular: St. Joseph's High School. Not just because of the competition, but because afterward, we would go to Pizza Hut. And coming from where I came from, Pizza Hut felt like a high-class restaurant.

I remember placing in the top ten in a lot of those competitions. People started telling me, "*You should be an engineer*," and "*You're good with numbers.*" And I believed them, because numbers made sense to me. They had rules. They had structure. And if you followed the process, you could always find the answer.

The Test I Never Fixed

I remember one test in particular. Algebra. Word problems, the kind I loved the most. I took the test, got it back: a 96.

But my teacher stopped me. She looked at my paper and said, "*Teddy... are you sure you don't want to take another look at this?*"

So I did. I looked over my work and saw it immediately. I had missed a step. Not the answer—the process. I knew exactly where I went wrong. I knew I could fix it.

And then I made a decision. I handed it back exactly the way it was. I never told her I saw the mistake. Because even then, I understood something I could not fully explain yet: mistakes are part of the process. And I had given it my best.

Later, in high school, I remember saying something during a news interview: "*Math is the only thing that is true to me.*" At the time, I did not fully understand what I meant. But I felt it. Because numbers do not lie. Even when we misunderstand them, they remain true.

The Problem in Front of Us Now

Now I am looking at a different kind of problem. Not on paper, not in a classroom, but in the real world. And just like back then, there are rules, there is structure, and there is an answer. But it is more complex, and it affects all of us. I call it the Big Three.

Problem One — The End of Labor

Robotics, automation, artificial intelligence. These are not just tools. They are replacements—removing wages, jobs, traditional income. So the equation becomes: if labor disappears, what replaces income? That is the unknown. That is the x.

Problem Two — The Weight of Debt

The United States is approaching $40 trillion in national debt. And debt is not just a number. It is an obligation. It requires someone to hold it, to buy it, to believe in it. So the equation becomes: if traditional buyers step back, who absorbs the debt? That is the second x.

Problem Three — The Weakening Dollar

The global system is shifting. Countries are exploring alternatives. Trade is evolving. The dominance of the U.S. dollar is being questioned. So the equation becomes: if the dollar weakens, what becomes the new financial layer? That is the third x.

The System That Begins to Solve It

Now step back and look at all three problems together. Labor disappears. Debt remains. Currency shifts. This is not three separate problems. It is one system under pressure. And when systems are under pressure, they evolve.

The New Model

1. Corporations Become Financial Engines

Not just businesses, but financial institutions. They introduce digital wallets, reward systems, financial layers inside their own platforms. But behind the scenes, they must hold something real: government debt, treasuries, bonds. They buy it, they hold it, they earn yield on it. And then they pass a portion of that yield back to people.

2. People Become Financial Participants

Every person with a smartphone now has access to a digital wallet, financial rewards, yield participation. Not because they worked, but because they are connected to the system. So now the government gets buyers for its debt, corporations earn a spread, and people receive value. It feels like a benefit. But it is actually a new structure.

3. Every App Becomes a Wallet

Airlines, retailers, tech platforms—all become financial distribution channels. Your phone becomes your bank, your wallet, your access point to the entire system. And now the system is everywhere.

4. Universal High Income Emerges

If people are no longer working, they still need income. So the system evolves into what I call Universal High Income, funded by automated production through AI and robotics, government debt markets, and corporate financial infrastructure. Income is no longer tied to labor. It is tied to participation in the system.

The Hidden Layer

But every system has structure. And every structure has control. Because if your income is digital, it can also be tracked, adjusted, restricted. The same system that can give can also limit. That is the part most people will not see at first.

The Multiplication Effect

Now take this model and multiply it. Not one corporation—all corporations. Every airline, every retailer, every tech platform, all holding government debt, all distributing value, all connected to the same system. Now scale that globally.

This is no longer an economy. This is an infrastructure of value distribution.

The Final Equation

Now go back to the three problems. Problem one, the loss of labor, is answered by Universal High Income. Problem two, the national debt, is answered by distribution through corporations and global participation. Problem three, the weakening dollar, is answered by a new digital monetary system—crypto, stablecoins, central bank digital currencies.

The equation begins to balance. But how it balances is what determines the future.

My Truth

I may not have every step perfect. Just like that test, there may be parts of this equation that I do not fully see yet. But I know this: I have looked at the problem, I have followed the structure, and I have given it my best. Because I see something forming. Something real. Something that needs to be understood.

And just like my teacher did back then, I am asking you to take a second look. Not at me. At the system.

Closing Line

Because in the end, this is not about being right or wrong. It is about seeing clearly. And once you see the equation clearly, you can decide where you stand in it.

Because clarity does not stay theoretical. It shows up in real moments, in real decisions, in the choices we make when no one is watching but everything is shifting. Once you see clearly, you do not just think differently. You move differently. You choose differently. You begin to recognize that certain moments are not random. They are positioned. They are aligned. They are calling you to pay attention.

And sometimes those moments do not look like turning points at first. They look like ordinary decisions—a trip, a conversation, a place you have been before but are now seeing for the first time with clarity.

And that is where this story begins. Not in theory. Not in systems. But in a moment that did not seem significant at first, until everything started to connect.

Chapter 28
THE TABLE, THE CHAIR, AND THE CALL

A Decision, Not Just a Trip

There are moments in life when everything begins to align. Not by accident, but with intention. This journey did not start as a chapter. It started as a decision: a decision to go to Buffalo Niagara International Airport, to catch a flight to Hartsfield-Jackson Atlanta International Airport, to see my daughter for her fifth birthday.

Weeks before that trip, I had gone to the DMV to get my Real ID. The rules had changed. Without a passport, a standard New York State license was no longer enough to travel. And at that moment, I could not find my passport. So I knew I might have to pay the price for that oversight. When my Real ID did not arrive in time, I had no choice but to go to the airport and figure it out.

When I got there, the lines were long. TSA was packed. People moving, waiting, frustrated—the kind of moment where everything feels rushed and unclear. But eventually, I made it through.

And what started as a simple trip became something more. Because that journey placed me somewhere I did not expect. In a new environment, around new energy. On my daughter's birthday, a moment of life, of renewal. And in that same space, I found myself in church, in a ceremony filled with presence, community, and purpose.

Without realizing it at the time, that was the beginning. A return to clarity. Because sometimes, before you can understand the world clearly, you have to quiet everything around you. And in that clarity, you begin to see what matters—not just what is changing, but what must be chosen. Because in the end, the future will not be defined by what was possible. It will be defined by what we understood in time to act.

And that moment, that experience, led me to something deeper—something that brought everything we have discussed into one place. A place where presence meets purpose, where history meets responsibility, where awareness meets decision. And that is where this journey takes us next.

Where It Began

It began in Marietta, Georgia. On February 15th, my youngest daughter's fifth birthday, I walked into Pleasant Grove Missionary Baptist Church with my wife and my daughters. It was my first time there. And that day, there was an Ordination Service taking place—a black-and-white ceremony about calling, character, and readiness to serve.

Without planning it, I was dressed for it: black pants, a white dress shirt. For a moment, everything went quiet, and one question stayed in my mind: "*Who comes here?*" I did not know what the ceremony was at the time, but I knew it was important. I am a Mason. I understand structure. I understand ceremony. And I could feel it.

So I leaned over to my wife and said, "*I feel like I'm part of this ceremony.*" At that moment, I did not fully understand why. But I felt something immediately: power, history, presence. This was not just a church. It was a living system, a community that had been built, protected, and carried forward through generations.

Ash Wednesday

Just days later, on Ash Wednesday, February 18th, we returned. This time, we drove into the church parking lot, and even from the car there was a process—a way to receive blessings. There was order, care, structure. Ashes were placed on my forehead in the shape of a cross: a reminder of repentance, mortality, humility. But also renewal.

The Power of Structure

Inside that same church, on February 15th, during the Ordination Ceremony, everyone participating was dressed in black and white. There was discipline in the room. Respect. Alignment. And even though I sat in my seat, I did not feel like an observer. I felt like I was part of it. Because real community does not just include you—it recognizes you.

And in that moment, even before I had the words for it, I began to understand something deeper: structure creates belonging, alignment creates clarity, community creates strength. What I was experiencing was only the beginning. Because real community does not isolate you. It pulls you in.

That church showed me something I will carry with me: you can be a community within a system without losing your soul.

The Shift

Then the journey continued. March 21st, Buffalo, New York. A different church, a different energy: Northgate. Younger, more diverse, different races, different backgrounds, different stories. But the same foundation: love.

That day, there was a conference called "Hearing the Father's Voice." And I walked forward. A minister spoke over me, not casually, not lightly, but with clarity.

The Vision

He said: "*You are sitting at the table of the Father, prepared to eat bread and wine. And a force comes and kicks your chair out from under you. It breaks. But you get up. You replace the chair. And you sit again. And it happens again. And this time, you make a decision: no more chair. You choose to eat from the floor. And when the force returns, you see it clearly. And because you see it, you destroy it. And in that moment, you realize what you were fighting was real. And because you can now see it, you can remove it—not just from your life, but from others.*"

The Alignment

And in that moment, everything connected. Because March 21st, 2026, was not just the day of the message. It was also my wife's and my sixth wedding anniversary. And that meant something deeper. The covenant still stands—through every challenge, through every test, through every moment of pressure. What was built was still intact. And now I could see it differently.

Because this was not just about a vision. It was about understanding—understanding that there are forces, patterns, and systems that operate whether we recognize them or not. But the moment you see clearly, everything changes. Because awareness is not passive. It is power. And that was the shift, not just in belief, but in sight. Once you can see, you can no longer move the same.

The Full Meaning

When I stepped back and looked at it all, I could see the pattern. Not random, not scattered. Aligned.

Life → Discipline → Reflection → Surrender → Revelation → Renewal

Not just a sequence of events, but a movement of awareness. Each stage revealing the next. Each moment building on the last. This was not just a timeline. This was transformation—a process of releasing what no longer aligned, and recognizing what had always been there. Understanding that I am not separate from what is unfolding. I am part of it, moving through it, becoming aware within it, and choosing, with clarity, how to rise.

Because resurrection is not just a moment. It is a decision. A decision to rise, to see differently, to think differently, to move with intention. To step fully into what you now understand.

And in that realization, it became clear: this was never just about me. This is a pattern, a message, a blueprint. The same way I was being called to rise with awareness, humanity is being given that same opportunity—to rise with understanding, to rise with clarity, to rise with purpose. Because the future is not something we simply enter. It is something we rise into.

What It Means

This entire journey unfolded while I was writing this book. Not outside of it, not separate from it. Within it. And that revealed something I could not ignore: this book is not just information. It is lived experience. Clarity earned through discipline. Awareness revealed through reflection. Understanding uncovered through alignment. Every page is not just written. It is a record of becoming.

Because understanding does not come from reading alone. It comes from living, observing, and becoming aware.

The Deeper Truth

We are living through a moment of transition. Systems are shifting. Ownership is evolving. And people are being pulled into change, often without fully understanding what is happening. But this journey revealed something essential: before we can understand systems, we must understand ourselves. Before we can participate in ownership, we must develop awareness.

Because without awareness, we do not shape systems. We are shaped by them. And that is the difference—between reacting and responding, between existing and participating, between being carried by the moment and rising within it.

The Realization

This journey was not random. It was alignment. And if you are here, reading this, then maybe this moment is not random for you either. Maybe this is your moment to pause, to reflect, to see more clearly, to recognize what has always been there, waiting to be understood.

Because in the end, this is not just about a book. It is about awareness. It is about responsibility. It is about choice. So the question is no longer, "What does this all mean?" The question is, "What will you do with what you now understand?"

Because the future is already unfolding. And the invitation is clear: rise with awareness, move with purpose, take your place in what comes next. Because this moment is not waiting. And neither should you.

Final Reflection

As I prepare to return to Northgate on Easter Sunday, my birthday, I know I am not returning the same person. Because something has shifted. Not around me, but within me. I understand now, clearly: this is not just about me. This is about my family, my community, my city, my generation, and the generations coming behind us.

I have done the fasting. I have done the reflection. I have sat in the silence. I have paid attention to the alignment. I have received the message. And now, it is time—to build, to teach, to connect, to lead. We have done the scouting. We understand the game plan. We see the field clearly now.

Clarity is no longer something I am searching for. It is something I am standing in. And from that clarity, one truth becomes undeniable: clarity does not arrive for comfort. It arrives for responsibility. Because once you see clearly, you begin to recognize that what shifted within you is not happening in isolation. It is connected to something much larger unfolding around you—the same patterns, the same pressure, the same quiet disruption moving through systems, structures, and institutions that once felt permanent. What became clear within is now becoming visible in the world.

And as that clarity expands, it reveals something most people are not yet prepared to face: we are not slowly transitioning from one system to another. We are standing in the space between them, where the old ways no longer fully hold, and the new ones are not yet fully understood. And in this space, something deeper is forming, quietly shaping how we live, how we move, how we are seen. Because the systems emerging now are not waiting to be explained. They are already being built.

So as you step forward, into community, into participation, into what comes next, remember this: you do not have to lose your soul.

Chapter 29
THE SYSTEM THAT SEES EVERYTHING

There is a model forming. Not slowly, not quietly, but clearly, for those willing to see it. This is not just a model of government. It is a model of control. A model of civilization. A model that is not asking for permission, because it is being built through adoption.

And at the center of this model is a belief: that the system knows more than the individual, that the collective is more important than awareness, that security is worth more than freedom, that conformity is the future.

The Architecture of Control

It does not begin with force. It begins with convenience. One language, one interface, one system of interaction. The smartphone becomes the universal translator. The internet becomes the common square. Artificial intelligence becomes the mortar binding everything together.

And rising above it all, a central tower. Not a physical tower, but a digital one—a system that sees everything, tracks everything, learns everything. Not through conversation, but through collection: facial recognition, fingerprints, iris scans, voice patterns, typing rhythms, browsing behavior. Every movement, every interaction, every decision—captured, stored, analyzed. Not to understand you as a human, but to define you as a profile.

None of this requires a single dramatic announcement. It arrives one convenient feature at a time: a faster checkout, a smoother login, a phone that already knows what you want before you finish typing it. Each step feels small. Together, they form the architecture.

The Profile Becomes the Person

And once the profile is complete, the system no longer asks who you are. It decides. Your habits become your identity. Your patterns become your permissions—where you can go, what you can buy, what you can access. All determined not by your voice, but by your data.

This is where currency changes. Money is no longer neutral. It becomes programmable. It can be turned on or off, restricted to certain uses, limited by time, directed by policy—a currency that can expire, a currency that can guide behavior, a currency that does not just move, but manages.

And when that currency connects to your profile, the system no longer needs enforcement. It has alignment.

The System That Replaces Labor

Now bring in AI. Bring in robotics. And remove labor—not most jobs, all of them. Production becomes automated. Services become automated. Decision-making becomes automated.

And now the question shifts: if people are no longer needed for production, how do they participate? This is where the system offers an answer: universal income, distributed digitally, managed centrally. But not freely given. Conditioned. Because when income flows through a programmable system, participation becomes compliance.

The Fully Integrated Life

Now connect it all: health records, financial behavior, mobility patterns, education history, communication networks, consumption habits, social positioning. All integrated, all synchronized, all real-time.

They call these smart cities. Cities that think, cities that respond, cities that optimize. But optimization for what? Efficiency? Security? Or control? Because in a system where everything is connected, nothing is separate. And in a system where nothing is separate, everything can be managed.

This Is Not Theory

This is not hypothetical. In China, elements of this system already exist: a growing integration of surveillance infrastructure, facial recognition used in public and commercial spaces, pilot programs and regional systems that evaluate trust, compliance, and behavior. Not one unified system everywhere, but enough to see the direction. A society where data is not separate from governance. It is governance.

At the same time, in the United States, new models are emerging in different forms. Starbase, an evolving company-led city concept connected to SpaceX, represents another version of centralized design, the same pattern already discussed earlier in this book with the return of the company town. Not identical, not the same intention, but part of the same question: what happens when infrastructure, technology, and governance begin to merge?

The Quiet Trade

This system does not arrive with a demand. It arrives with an offer: convenience, speed, safety, efficiency. And in exchange, it asks for something subtle. Not money. Not even obedience. Just awareness. Because once awareness is traded away, everything else follows.

Do You See It Now

This is not about fear. It is about clarity. Because there are always two paths. One where technology expands awareness and empowers individuals to participate in ownership. And one where technology replaces awareness and systems begin to decide on behalf of humanity.

Both paths use the same tools: AI, data, infrastructure, digital systems. The difference is not the technology. The difference is who holds the authority. And whether people remain participants, or become profiles. This is the same divide raised earlier in this book between structured ownership and controlled systems, just seen now from the surveillance side of the equation rather than the financial one.

Final Line

This is not a warning. This is recognition. Because once you can see the system clearly, you can no longer pretend you are outside of it. And from that moment forward, you are no longer just living inside the future. You are deciding your position within it, the same way every chapter of this book has been an invitation to decide your position rather than simply observe.

Because what comes next will not be defined by what is announced, but by what is felt. A shift beneath the surface. A quiet unraveling of what once felt certain—not breaking all at once, but loosening, until it no longer holds.

And in that moment, you begin to notice it. And once you begin to notice it, you start to see something else. Not just how systems operate, but who is building them. Because the future is not forming on its own. It is being designed, layer by layer, decision by decision, by individuals who understand how systems work and where they are going.

And when you see that clearly, the question shifts again. From "What is happening?" to "Who is shaping it?" And more importantly, how are they doing it?

Chapter 30
OWNERSHIP AT SCALE

There are people who build companies. And then there are people who build systems that reshape entire industries. Elon Musk is one of those people—not because of any single company, but because of how all of them connect.

The Illusion of Separation

Most people see his businesses as individual ventures: Tesla, SpaceX, Starlink, Neuralink, The Boring Company, xAI, X. Different names, different missions, different headlines. But that is the surface. When you look deeper, the separation disappears. These are not independent companies. They are coordinated layers of a larger system.

Movement, infrastructure, connectivity, intelligence, communication, expansion, integration—each layer solving a different problem, but all moving toward the same direction. Not competing. Aligning. Not isolated. Reinforcing. This is not random innovation. This is intentional design.

What Power Actually Looks Like on the Ground

I learned this long before I ever studied systems at scale. I learned it working on the 296 Monroe project. And it taught me something most people do not see until they are inside of it: nothing gets built—not land, not buildings, not permits, not construction—without alignment with political power. That is not theory. That is reality.

There were countless nights in block club meetings, listening, explaining, building trust. Not because the idea was not strong, but because support had to be structured. Before anything is approved, there has to be agreement, there has to be alignment, there has to be backing. And that backing flows through people who sit in positions of decision.

That experience revealed something deeper: power is not just about what is built. It is about who has the authority to allow it to exist in the first place.

And that is why this moment matters. Because as systems scale, and as technology companies begin to operate at levels once reserved for governments, a new question emerges: who holds them accountable? Because if platforms become infrastructure, applications become financial systems, and individuals begin to hold percentages of the assets humanity depends on, then power is shifting. Not disappearing. Shifting.

The Missing Balance

We have always had systems of checks and balances: local government, state government, federal government. But what happens when influence moves beyond those boundaries? What happens when the systems shaping daily life are no longer fully governed by traditional structures?

We may be approaching a moment where regulation is not optional, where policies must evolve, where leadership at every level must begin to ask a harder question: how do we ensure that the systems being built still answer to humanity?

The Real Tension

This is not about stopping innovation. It is about understanding power, and making sure that as ownership expands, accountability expands with it. Without that, the same pattern repeats: value is created by many, control is held by few. And the system, no matter how advanced, begins to drift away from the people it was meant to serve.

This Is Not Business—This Is Architecture

Most businesses are built to operate within systems. This is different. This is building the system itself. When you design how people move, how they connect, how they communicate, how intelligence flows, how infrastructure scales, you are no longer participating in the future. You are structuring it. This is architecture, at scale.

Beyond Companies — The Environment

And then there is something else. Not a company, not a product, but a signal: Starbase. At first glance, it looks like a launch site. But when you step back, it becomes something different—a physical environment designed around a system. Work is centralized. Infrastructure is controlled. Daily life begins to align with the mission itself.

This is not just real estate. This is system extension into the physical world. Because at a certain level, ownership is no longer just about companies. It becomes about environments—the places where systems operate, scale, and sustain themselves.

Ownership Beyond Equity

Most people have been taught that ownership means shares, equity, control. But that is a limited definition. At this level, ownership is not about holding pieces. It is about holding position—position inside the systems that create value: the networks, the platforms, the infrastructure, the environments. Because once dependency is created, participation follows.

The Alignment Most Never See

This is why it feels powerful—not just because of innovation, but because of alignment. Each company strengthens the next. Each layer feeds the system. And now even the environment begins to support the system itself. Nothing wasted. Nothing disconnected. That is not coincidence. That is design.

The Gap Between Players and Architects

Most people are trying to get a job, build a business, find stability, inside systems they do not fully understand. They learn the rules. They follow the structure. They try to win the game as it was given to them.

But there is another level, a level most never reach, because they were never taught to look for it. The question changes from "How do I succeed here?" to "Who designed this?" And then again, "Where is this going?" Because once you ask that, you are no longer just playing. You are observing the system as it is being built.

The Shift

And that is where everything changes. Because now the goal is no longer success within the system. It becomes positioning within its creation. Not reacting to opportunity, but recognizing where opportunity originates. That is ownership.

The Real Lesson

This is not about Elon Musk. It never was. It is about awareness. Because once you see this clearly, you realize something deeper: the future is not just being created through technology. It is being structured through ownership—layer by layer, system by system, environment by environment. And those who understand that early do not just adapt to change. They align with it. They build within it. They help define it.

Closing

Most people will experience the future through systems they never questioned. A few will understand it. And an even smaller group will recognize the pattern early enough to position themselves before it becomes obvious. Because once it becomes obvious, the system is already in place. And by then, you are no longer choosing it. You are living inside it.

This is where the shift reveals itself. Not in headlines, not in announcements, but in what begins to quietly give way. Because when a system is no longer questioned, it does not disappear all at once. It loosens. It weakens. It fades beneath the surface until what once felt certain no longer holds the same weight. And for those who were not looking, it feels sudden.

But for those paying attention, this moment was always coming—the point where the old structure can no longer carry what is emerging.

And in that space, between what is ending and what is not yet fully understood, there is a shift you do not hear. You feel it.

Chapter 31
THE SPACE BETWEEN SYSTEMS

Where Humanity Is Measured

There is a silence coming. Not the kind you hear—the kind you feel. A slowing, a thinning of what once felt certain. A quiet realization that something familiar is no longer holding.

No announcement. No moment where the world stops and says, this is where everything changed. It will happen gradually, and then all at once. The systems people trusted—work, income, structure, identity—will begin to loosen their grip. Not because they failed. But because something more efficient replaced them. And efficiency does not ask permission.

When Work Disappears

There is a truth most people are not ready to face: work was never just about money. It was about meaning. It told people who they were, where they belonged, why they mattered. Take that away, and what remains? Not freedom. Not at first. At first, emptiness.

Because when structure disappears before understanding arrives, people do not expand. They drift. And drifting becomes anxiety, confusion, comparison, desperation. Not because people are weak, but because they were never prepared for a world where survival was no longer the organizing principle.

The Dangerous Middle

This is the part no one talks about. Between the old system ending and the new one being understood, there is a gap—a space where people do not know how to earn, do not know how to contribute, do not know how to position themselves, but still need to live.

That space is where instability is born. Not from poverty alone, but from disconnection. Because when people feel unseen, unneeded, uncertain, they do not just struggle. They begin to question everything. And when enough people reach that point, society does not just shift. It fractures.

Power Will Not Stay Empty

History has shown this again and again: power never sits still. When there is a vacuum, something fills it. And in this moment, what fills it will be shaped by the tools we are building right now. Not later. Now.

The Pulse

And in the middle of this transition, there will be a moment. A pulse. A brief window where everything feels unstable at once. Not random. Aligned. Pressure rising across systems: energy disruption, financial instability, supply strain. Moments that feel like crisis, because they are experienced as crisis. And in that moment, it will feel like everything is breaking at the same time.

This is where awareness matters most. Because history shows something clearly: in uncertainty, people do not just react. They accept. They look for stability, for direction, for something, anything, that restores order. And that is where decisions are made. Not slowly. Instantly, under pressure.

This may be one of the most important pulses in human history, because what is chosen in that moment does not just stabilize the system. It defines it. Will people drift into disconnection, fear, and social unrest? Or will they pause, see clearly, move with awareness, and choose to build something that includes everyone?

Because in moments like this, the future is not debated. It is decided. And if awareness is absent, systems will be accepted without understanding. Control will be embraced as safety. Structure will be rebuilt without ever asking who it truly serves. But if awareness is present, then even in instability, something else becomes possible. Not reaction. Choice.

The Two Uses of Intelligence

Artificial intelligence will not decide the future. But it will amplify whoever understands it first. And that creates two very real paths.

Control Without Resistance. Not force, not visible oppression, but something far more efficient. A world where access is conditional, behavior is tracked, and decisions are influenced before they are ever made. Where money can be restricted, movement can be guided, choices can be narrowed—not through punishment, but through design. So seamless that you do not even realize you are inside it. Because control in the next era will not feel like control. It will feel like convenience, safety, optimization. And that is what makes it powerful.

Awareness Without Permission. There is another path. Harder, less controlled, more demanding. A world where people understand the systems they use, ownership is visible and shared, and value is not hidden behind institutions. Where communities participate, individuals are positioned, and access is created rather than granted. Not perfect. But aligned. Because awareness requires responsibility, and most people have never been trained to carry it.

The Moment Everything Changes

There is a threshold approaching, whether acknowledged or not: the moment where intelligence no longer belongs only to humans. Where machines learn faster, adapt quicker, decide more efficiently. Some call it singularity. But the name is not the point.

The point is this: if intelligence surpasses humanity, and humanity does not understand the systems guiding it, then control will not need to be forced. It will be accepted. Because people will trade understanding for ease, awareness for comfort, responsibility for convenience. And slowly, without realizing it, they will hand over the very thing that once defined them: agency.

The Illusion of Safety

There will be a moment where everything feels stable again. Systems working. Needs met. Life, manageable. And many will say, we made it.

But beneath that stability, a question remains: who is actually in control? Not visibly, not politically, but structurally. Because once systems are in place, they do not need to control loudly. They control quietly—through defaults, through incentives, through limits no one questions. And by the time it is seen clearly, it is already normal.

This Is the Real Risk

The risk is not technology. The risk is entering a new world with an old level of awareness—operating inside systems you do not understand, believing you are free without seeing what is structuring your choices. That is where control lives. Not in force. In invisibility.

The Opening

And yet, this moment is not only risk. It is possibility. Because the same space that creates confusion can also create awakening. A reset. A chance to ask what was never asked: what is value? What is ownership? What is participation? What does it mean to live, when survival is no longer required?

The William S. Jordan Class

There are moments in life that do not just mark time. They mark transformation. The William S. Jordan Class was one of those moments. Not just a class, not just a process, but a passage—a crossing, a moment where something within you quietly shifts, and you realize you are no longer standing where you once were.

My cabletow brother always says, "*Mastercraftsman 110 is the best lodge in the world.*" At first, that might sound like pride. But when you experience it, you understand it is not about comparison. It is about standard. Because what is built there is not just knowledge. It is discipline. It is accountability. It is awareness. It is the reminder that becoming a man is not automatic. It is intentional.

And what I came to understand is this: you are not given the title of Master Mason. You grow into it. You carry it. You are measured by it—in how you think, how you move, how you show up when no one is watching.

The Moment I Understood

There was a moment, not inside the lodge, but in a church in Atlanta, at the ordination ceremony. And something in me became still. Not silent. Clear. It was not explained. It was not taught. It was felt: a recognition that what I had been walking toward had already been placed within me. That responsibility was not something being given. It was something being revealed.

And in that moment, I understood. Not with words. With awareness. That becoming a Master Mason was never about the ritual alone. It was about the alignment—the responsibility to carry yourself differently, to see differently, to lead differently, to live with intention, not reaction.

What It Really Means

Because the truth is, titles do not change people. Awareness does. Responsibility does. And once you feel that shift, you cannot go back to moving the same way. This is not about status. This is about standard. Not about being seen, but about seeing clearly. Not about power, but about discipline. Not about knowledge, but about how you apply it when it matters most.

The William S. Jordan Class was not the moment I became something new. It was the moment I recognized what I was always meant to carry. And from that day forward, I no longer moved just as myself. I moved with awareness. I moved with responsibility. I moved as a Master Mason.

Some information is not hidden. It is held. And to access it, you do not search for it. You step into it.

Closing

This is not the end of a system. This is the exposure of one. And in that exposure, a decision is waiting. Not for governments. Not for institutions. For individuals. Because in the space between systems, you are not just witnessing the future. You are deciding whether you will understand it, or be positioned by it.

Two paths. One shaped by awareness, alignment, and intentional ownership. The other shaped by confusion, passive participation, and systems chosen without understanding. Both will move forward. But they will not lead to the same outcome.

And this is where everything changes. Not later. Now. Because once you see the system clearly, the question is no longer about the future. It becomes personal. It becomes present. It becomes this: where do you stand within it? And more importantly, how do you move?

Chapter 32
THE PIECE ON THE BOARD

There was a question I was asked by someone close to me: "*If you were a piece on a chessboard, what piece would you be?*"

At first, it felt like a simple question, the kind of thing you answer quickly at a dinner table and move on from. But the more I sat with it, the more I realized it was not simple at all. Because that question is really asking something deeper: how do you see the game? And more importantly, how do you see yourself within it?

The Queen

I sat with that question overnight. It stayed with me, not lightly, but deeply. And the next day, I called him. I said, "*I've thought about it. I think I'm the Queen.*"

He paused. Then he said, "*You don't want people to identify you with something feminine, do you?*" And then added, "*What do you think you are, a pawn?*"

I smiled. Not because I was unsure, but because I was clear. My answer was never about gender. It was about perspective, and about how differently the two of us had learned to read the same board.

If I had to answer honestly, the closest piece would be the Queen. Not because of power, but because of range. The Queen moves in all directions. She is not limited to one path, not confined to one role. She sees space differently, and she moves accordingly.

That is how I have always operated—moving between community, business, sports, education, and vision. Not staying in one lane, but understanding how they all connect. Because most people play the game from where they stand. They move based on what is in front of them. But some people see the whole board.

But That's Not the Full Truth

Because over time, I realized something else. I am not just moving pieces. I am studying the board—questioning the structure, looking at how the game is designed in the first place. Why certain pieces move the way they do. Why some positions feel limited. Why some players never get the same opportunities, no matter how well they learn to play the pieces they were handed.

That is when the understanding shifted. This is not just about playing the game.

It's About Understanding the System

Most players react, protect, compete. They try to win within the rules they were given. But there is another layer, a deeper level, where you begin to ask: who created the rules? What defines value in this game? And what happens if the structure itself changes? That is not just strategy. That is awareness. It is the same question this entire book has been circling, just asked now in a language simple enough for a chessboard to hold.

Why I Felt Stuck

There were times in my life when I felt out of place. Frustrated. Misunderstood. Starting big visions but not always finishing them. And now I understand why. I was thinking like a Queen, but operating inside systems designed for pawns—systems that reward repetition, not vision; structure, not awareness.

And when you see more than the system is designed to hold, it creates tension. Not because something is wrong with you. But because you are seeing beyond what is expected, in rooms built for people who were never asked to see that far.

The Shift

But something changed. I began to understand that seeing the board is not enough. Vision without structure does not move anything. Awareness without execution does not build anything. I had spent years able to see five moves ahead while still fumbling the very next one, and for a long time I mistook that gap for a personal failure rather than a missing skill.

So the shift became clear: not just seeing everything at once, but learning how to move one square at a time, with intention, with discipline, with clarity. That is when everything changes, because now the same perspective that once felt overwhelming becomes precise. The board does not get smaller. Your ability to move across it gets sharper.

Beyond the Board

And then the truth reveals itself: you are not just a piece on the board. You are not just playing the game. You are part of something deeper. Because once you understand the system, you begin to see that the board itself can change. That the rules are not fixed forever. That structure is something that can be built, not just followed.

That is where real power lives. Not in control. In awareness. The same awareness this book has been asking readers to build in every prior chapter, whether the board was a nonprofit, a city, a currency, or a company town.

The Real Question

So the question was never, what piece are you? The real question is: are you trying to win the current game, or are you here to build a new one? Because everything I have written, everything I have built, points to one truth. I am not here just to play. I am here to change how the game is understood, for the next person who inherits this same board without ever being shown how to read it.

Closing

And once you see the board clearly, you realize something most never do: you were never just a piece. You were always part of the design. And once you understand that, the question changes again. It is no longer about position. It is no longer about movement. It becomes about direction.

Because a system can only take you so far before a choice must be made. Not unconsciously. Not by default. But intentionally. That is the same lesson that has run through every chapter of this book, whether it was Buffalo, a nonprofit, a community, or a company town: the moment you stop reacting to the board and start studying it is the moment ownership becomes possible.

And that is where everything converges. Not in theory. Not in observation. But in decision.

FROM PLAYING THE GAME
TO MASTERING THE SYSTEM
TO BUILDING OWNERSHIP
SEE THE BOARD. STUDY THE SYSTEM. OWN THE FUTURE.
CHESS
THE VISIBLE GAME
PIECES
Every player has a role.
MOVES
Movement is limited by the rules.
STRATEGY
Make the best move to win.
WIN / LOSE THINKING
Focus on beating the opponent.
Most people stop here. They play the game.
SYSTEM
THE HIDDEN STRUCTURE
WHO CREATED THE RULES?
Every system is designed by someone for a purpose.
HOW VALUE IS DEFINED
Value is not natural, it is determined.
WHO CONTROLS MOVEMENT?
Access, resources, and opportunities are controlled.
INCENTIVES & CONSTRAINTS
The system rewards certain behavior and limits the rest.
Few people get here. They study the game.
OWNERSHIP
THE LEVEL OF AWARENESS
UNDERSTAND THE SYSTEM
Clarity gives you power.
POSITION WITH INTENTION
Align your purpose with the right opportunities.
CREATE NEW PATHWAYS
Build solutions that don't exist yet.
BUILD, NOT JUST PLAY
Shift from participant to architect.
The rare few reach here. They build a new game.
THE MINDSET SHIFT
AWARENESS
See beyond what's in front of you.
KNOWLEDGE
Understand how the system works.
OWNERSHIP
Take action. Create impact. Change the game.
YOU WERE NEVER JUST A PIECE.
YOU WERE ALWAYS PART OF THE DESIGN.

Chapter 33
TWO PATHS, 1 FUTURE

A Decision Point for Humanity

Every chapter in this book has been building toward this one moment, whether it was Buffalo losing its industrial center, a nonprofit trying to stabilize lives inside an unstable structure, a billionaire quietly buying land for a company town, or a football field where the outcome was decided before the ball was ever snapped. Different settings, different stakes, but always the same underlying question: do you see the system clearly enough to choose your place within it, or does the system simply choose for you, quietly, without ever asking your permission?

I have watched this moment approach. Not suddenly, but steadily. There comes a moment when the future must be chosen, not assumed—where direction is no longer assumed, but chosen. This is that moment.

What once felt distant is now present. What once felt theoretical is now unfolding. Throughout this journey, you have seen what is changing. You have examined systems under pressure. You have witnessed the evolution of ownership. You have recognized the emergence of new infrastructure already shaping the world around you. You have seen how this touches communities, businesses, athletes, creators—not as separate stories, but as one connected shift.

And now we arrive at something deeper. Because information is not enough. Understanding is not enough. There comes a point where awareness must become decision. And that point is now, not at the end of this book, but at the end of this sentence.

The First Path — Movement Without Awareness

There is a path where people continue moving without fully seeing, where decisions are shaped by noise, urgency, pressure, fear. On this path, systems continue to evolve, but understanding does not expand with them. Participation becomes limited. Ownership becomes concentrated. Access becomes controlled.

This is the path Buffalo walked once already, when the factories closed and an entire generation was left to adjust after the fact, without ever being shown the new map. It is the path of the nonprofit that keeps chasing the next grant cycle without ever asking why the funding was structured that way to begin with. It is the path of working inside a system for decades and only recognizing, much later, that the system was never designed to let you own anything—only to let you participate.

And over time, people find themselves inside structures they never fully understood—not because they lacked ability, but because they never paused to see clearly. They adjust. They react. They follow. And slowly, what feels like participation becomes positioning without control.

The Second Path — Movement With Awareness

There is another path, grounded in awareness, where individuals, communities, and organizations choose to understand what is changing before they are forced to respond to it. On this path, people move with intention. They ask better questions. They seek clarity before action.

They position themselves, not perfectly, but consciously. They do not wait for certainty. They move with understanding.

This is the path of the Community Ownership Framework discussed earlier in this book: awareness first, then understanding, then trust, then action, then ownership, one deliberate step at a time rather than one dramatic leap. It is the path of the young person leaving foster care who is taught not just to save, but to understand how ownership actually compounds. It is the path of a community that stops asking what it lacks and starts asking what it already holds, and how to structure it.

And that changes everything. Because this path does not remove uncertainty. It does not guarantee ease. But it creates something far more valuable: choice. Awareness expands what is possible. And choice is where ownership begins.

The Future Is Already Being Built

The future is not forming in one place. It is not controlled by one system. It is being shaped every day, across industries, across communities, across conversations happening in real time. And whether it is recognized or not, you are already part of that process. Not later. Now.

The question is not whether you will participate. You already are. The question is: how will you show up within it? Will you move without understanding and adjust when forced to? Or will you take the time now to see clearly, to think deeply, to act with intention? That question does not belong only to nations, corporations, or communities. It belongs to you, in whatever room you are standing in as you read this.

The Moment That Will Be Remembered

There will come a time when this moment is no longer uncertain, when what feels complex now becomes obvious, when the direction of change is clearly defined. And when that day comes, this moment will matter. Because this is where awareness begins. This is where people decide how they think, how they respond, how they choose. History rarely marks these moments while they are happening; it marks them afterward, looking back at who paused long enough to see clearly and who simply kept moving.

The Invitation

So wherever you stand right now, this is your moment. Not to rush, not to react, but to pause, to reflect, to become aware. Not out of fear, but out of understanding. Because the future is not asking for perfection. It is asking for participation.

And in a world moving faster than ever, your greatest advantage is not speed. It is clarity. Because clarity creates direction. Direction creates action. And action, when aligned with awareness, creates ownership. And ownership is where the future begins.

Two paths. One future. The choice was never whether the future would arrive. It was always whether you would meet it awake or asleep, positioned or merely present. Everything in this book has been an attempt to hand you the map before you needed it, rather than after. What you do with it now is the only part of this story left to write, and it is the only part I was never able to write for you.

CLOSING REFLECTION

Nothing was ever hidden—only unseen. Awareness was always the doorway. Now that you see, you cannot return to not knowing.

With that awareness comes responsibility—not to resist, but to build.

The future was never separate from you. It has always been shaped by what you choose to see, and who you choose to become.

– Theodore McDuffie

FINAL MESSAGE — TO HUMANITY

I have watched you build. I have watched you struggle. I have watched you rise, fall, and rise again. I have watched you create systems, and then search for your place within them.

And now you stand at a threshold. Not of endings, but of awareness.

This moment is not asking you to be perfect. It is asking you to be present. Because everything that comes next will not be decided only by systems. It will be shaped by people who are willing to see clearly. You are not separate from the future. You are part of it. Not later. Now.

And the invitation is clear: rise with awareness. See what is changing. Understand what is being built. Recognize your position within it. Not from fear, but from clarity. Move with purpose. Do not drift. Do not react without understanding. Choose your steps. Ask better questions. Align your actions with intention. Because movement without awareness leads nowhere.

Take your place in what comes next. Not on the outside, not waiting, but inside the process—learning, building, participating. Because the future is not reserved for a few. It is shaped by those who engage with it. And humanity has never needed perfection to move forward. Only awareness. Only courage. Only the willingness to step into the unknown with clarity.

There will always be change. There will always be uncertainty. There will always be systems evolving beyond what is immediately understood. But there will also always be something greater: the ability to learn, the ability to adapt, the ability to come together and build something better.

That is where your power lives. Not in control of everything, but in awareness of what matters. Not in knowing every answer, but in choosing to understand. And when enough people choose awareness, something shifts—not just in systems, but in humanity itself. Alignment replaces confusion. Connection replaces division. Purpose replaces reaction. And from that place, a different future becomes possible. Not one built in isolation. One built together.

So this is not an ending. It is a call. A moment to rise. A moment to move. A moment to take your place. Because what comes next is not waiting to be discovered. It is waiting to be built.

And it begins with awareness.

AUTHOR REFLECTION

If you have reached this point in the book, I want to thank you. Not simply for reading these pages, but for taking the time to truly think about the ideas explored within them. In a world that moves quickly, taking time to think is not common. And yet, it is where awareness begins.

This book was never meant to present a final answer about the future of financial systems, technology, or ownership. The truth is, none of us can predict exactly how the coming decades will unfold. Technology will continue to evolve. Economic systems will continue to change. New opportunities will emerge while old systems adapt or disappear. Many of the structures and institutions societies have relied on for decades are now being questioned, and at the same time, new systems are beginning to take shape, though many of them are not fully formed. That creates something we can all feel: uncertainty.

Uncertainty can feel uncomfortable. But history teaches us something important: moments of uncertainty are also moments of creation. When existing systems begin to shift, space opens—space for new ideas, new models, new ways of organizing value, cooperation, and responsibility. And in those moments, it is not only institutions that shape what comes next. It is people.

Throughout history, some of the most meaningful change has come from individuals and communities choosing to act with awareness: families supporting one another, communities building solutions together, small networks sharing knowledge, resources, and trust. These quiet forms of cooperation often become the foundation for the systems that follow.

Which is why this moment asks something important of each of us. It asks us to pause and ask a sincere question: what kind of world do I want to live in? A world guided by fear, or by understanding? A world shaped by division, or by cooperation? A world where people wait, or a world where people take responsibility?

The answers to those questions will not come from institutions alone. They will come from people—from everyday choices, from conversations within families, from communities deciding what they stand for, from individuals choosing awareness instead of distraction, curiosity instead of fear, responsibility instead of avoidance.

There are still many questions ahead, and there is still a great deal of future to be lived. This book was written in the spirit of that awareness. Not from fear, but from a belief in humanity's ability to learn, grow, and build systems that reflect its highest values.

Throughout my life, as a coach, as a father, and as someone who has spent years trying to understand the systems shaping our society, I have come to believe something simple: the future will not belong only to the most powerful institutions. It will belong to those who are willing to learn, those who remain curious, those who take responsibility for understanding the systems shaping their lives, and those who choose to participate with awareness.

Because ownership, at its deepest level, is not only about assets. It is about responsibility—responsibility for the knowledge we seek, responsibility for the choices we make, responsibility for the future we help create for those who come after us.

That responsibility is something I think about often, especially when I think about my children. Because in the end, everything we build, everything we learn, and everything we pass forward is not only about us. It is about them.

And that is why the next words in this book are not written for the world. They are written for them.

A LETTER TO MY CHILDREN

AJ, Trevor, Ava, and Nia,

If you are reading this, then you have reached the end of something that was never just a book. It was a journey. A process of seeing. A process of becoming aware. And in many ways, a process of becoming.

Everything I studied, everything I questioned, everything I struggled to understand, was never only for me. It was always for you.

There were days I sat alone, trying to make sense of systems I was never taught to see clearly. Long nights. Quiet moments. Questions without immediate answers. Times when the path did not feel certain. But even in those moments, there was always something guiding me forward—a deeper understanding that this work mattered. Not just for where I was, but for where you would one day stand.

Because as I moved through life, something became clear to me: many people are taught how to participate in the world, but very few are taught how to understand it. And once you begin to understand, everything changes.

AJ and Trevor, you saw the early stages. You saw me learning in real time. You saw me searching. You saw me trying to understand things I was never given. I hope you carry this with you: growth is not about certainty. It is about continuing forward, even when clarity has not fully arrived.

Ava and Nia, you came into my life at a time when my awareness had deepened. You reminded me that the future is not abstract. It is personal. It is real. It is alive in the people we love.

Each of you carries something powerful, something that cannot be measured, something that cannot be replaced. My hope is that you learn to recognize it, to trust it, to build from it.

I want you to understand ownership, not only as something external, but as something internal. Own your thoughts. Own your growth. Own your direction. Because ownership begins within, before it ever shows up in the world around you.

There will be moments of uncertainty, moments where the path is not clear. I have lived through those moments, and I need you to understand this: those moments are not obstacles. They are invitations—invitations to grow, to see deeper, to become more aware of who you are.

Never be afraid to ask questions. Never stop learning. And never believe that anything placed in front of you is beyond your ability to understand. Because everything you see in this world, every system, every structure, every opportunity, was created through awareness. And that same awareness lives within you.

You come from strength. You come from resilience. You come from intention. And you carry forward everything that was poured into you, not as weight, but as foundation.

And if you ever find yourself questioning whether you are ready, whether you are capable, whether you are enough, I want you to remember this: you are not separate from what is unfolding. You are part of it. You are shaping it, with every thought, every decision, every step forward.

And in that realization, something becomes clear: the systems we study, the value we create, the future we build, are not separate from us. They are expressions of awareness, moving, learning, and becoming through us.

So wherever life takes you, move with clarity. Move with intention. Move with awareness. And never forget, you are deeply loved. I am proud of you. Always.

With all my love,

Dad

AUTHOR'S LEGACY

If you are reading this long after my time, and you find yourself asking who I was, understand this first: I was not just a man trying to explain systems. I was a man trying to help people see clearly.

I lived in a time of rapid change—a time when technology was advancing, systems were evolving, and the future was being shaped in real time. But what I recognized early was this: the greatest gap was not access. It was not talent. It was not even opportunity. It was awareness.

So I dedicated my life to helping people understand how value is created, how systems are structured, how decisions shape outcomes. Not so they could fear the future, but so they could participate in it.

In my earlier days, I was a numbers guy. I enjoyed math. I found truth in it, because numbers do not lie. Even when we misunderstand them, they remain true.

I was also an American football player—a student of the game, a competitor, a coach. For more than four decades, I played it, studied it, and taught it. And what the game gave me was more than discipline. It gave me perspective. It showed me how systems work, how roles connect, how preparation meets opportunity.

But my life was never just about numbers or sports. I served humanity first. My journey began working with foster care youth, single parents, and families navigating instability. Later, I worked with individuals experiencing homelessness, walking alongside people in moments where life had stripped everything down to its most honest form.

And through all of that, I learned something most people miss: value is not always where the world tells you to look.

I have had friends that the world would call celebrities, and I have had friends that the world would overlook completely. But the reason we were connected was never status. It was never position. It was always measured by one thing: the love in their hearts.

Because I love humanity, fully. Even knowing that love comes with risk. Even knowing that sometimes that love will be met with misunderstanding, disappointment, or pain.

My heart has been broken more than once. But what I learned is this: if you keep your heart open—not guarded, not closed, not hardened, but open in love—then right around the corner, the most amazing love is always waiting.

That belief stayed with me. It shaped how I saw people, how I saw systems, how I saw the future.

Because I came from community, and I stayed connected to it—youth, families, people searching for direction, opportunity, and stability. I saw firsthand what happens when people are expected to operate inside systems they were never taught to understand. And I knew something had to change.

I was not here to impress people with knowledge. I was here to translate it—to make it real, to make it usable, to make it meaningful. Because understanding creates confidence. Confidence creates better decisions. And better decisions shape the future.

If you study my work, you will see a pattern: I believed that most division in this world comes from misunderstanding. That people can stand in the same moment, on different sides, looking at the same reality, and see something completely different. And that awareness, true awareness, has the power to bring clarity where there is confusion, and connection where there is separation.

I believed in responsibility. Not just personal responsibility, but collective responsibility—to family, to community, to humanity. Because the systems we build today become the environments future generations must live in.

And if there is one thing I would want remembered, it is this: ownership is not just about what you have. It is about how you think, how you choose, how you show up in the moments that matter most.

I understood something deeper as well, something that guided me beyond systems, beyond titles, beyond any single role I held: I was not separate from the world, but a part of it. Not observing change, but participating in it. Not disconnected from others, but connected through shared experience, shared responsibility, and shared possibility.

So if you carry anything forward from my life, let it be this: see clearly, think deeply, act with intention. And when the moment comes, because it always does, choose not from fear, but from understanding.

The future is not something we inherit. It is something we choose. Together.

ACKNOWLEDGMENTS

This journey did not begin with a book. It began with questions, with observation, with a desire to understand the systems shaping the world and my place within them.

Over time, as I studied more deeply, I came to realize something that changed everything: the greatest lessons I have ever learned did not come from theory. They came from faith, family, and football. Three forces that shaped me, not just in moments of success, but in moments of uncertainty. When things felt heavy, when direction was not clear, when life demanded more than I understood at the time, I leaned into faith, I stayed grounded in family, and I carried the discipline football taught me. Faith gave me perspective. Family gave me purpose. Football gave me structure. Together, they gave me the foundation to keep moving forward.

I want to first acknowledge my parents, who are no longer here physically, but whose presence lives in everything I do. They taught me something deeper than success, something deeper than ownership as the world often defines it. They taught me how to love, how to sacrifice, how to show up for others consistently, without conditions. They showed me that the greatest form of ownership is not in what you have, but in how you care for people. Mom, I wrote a book. And I hope that one day it helps someone, even if it is just one person.

I want to thank every community that embraced me and helped shape who I am—from neighborhoods to football fields, to classrooms, to workplaces, to spaces of faith. Each one gave me something. Each one expanded my understanding of people, of systems, and of life.

To my friends and extended family: thank you for your support, for your presence, for showing up in moments when it mattered most. Not always loudly, but consistently. And that consistency meant more than words can fully express.

To my brothers, my sister, and their children: thank you for your love and for allowing space for growth. Life is a journey, and we do not always move at the same pace. But your patience, your understanding, your presence, has meant everything.

To the mother of my sons: thank you for your dedication and commitment to our boys, and for the friendship and support that continues to shape their lives.

To my children: there are no words strong enough. I love you deeply. You teach me, every day, what truly matters. You are both my responsibility and my reflection.

And to my wife: thank you, for the lessons, for the perspective, for the growth, for the love. You helped me understand something that changed everything: the difference between reaction and awareness. You showed me, through experience, not just words, that who we are is not defined by what we say, but by what we consistently choose to do. And through that understanding, something deeper formed within me—a clarity, a recognition, a return to self.

Because in the end, this journey was never just about systems. It was about understanding myself within them. And through each of you, in different ways, at different times, that understanding became possible.

Thank you.

NEXT STEPS

Website: Empire Nexus

If this book brought you clarity, the next step is positioning. Because understanding the system is one thing. Knowing how to move within it is everything.

Email / Contact:

info@empire-nexus.com

Website:

https://www.empire-nexus.com/

THE COMMUNITY WEALTH ECOSYSTEM

EMPOWERMENT THROUGH EDUCATION

Establish local skills workshops for sustainable business models, digital literacy, and community leadership. Certify knowledge on-chain to create tangible value for members.

DATA-DRIVEN DECISION MAKING

Utilize predictive analysis and community-driven metrics to optimize resource use. Create transparent performance dashboards for all members.

GOVERNANCE TECHNOLOGY

Deploy transparent decentralized voting platforms and DAO structures to manage shared assets. Utilize smart contracts for fair resource allocation and dispute resolution.

FUTURE INNOVATION LABS

Fund internal R&D for next-gen green tech and community wellness initiatives. Pilot new tokenized utility models for community services.

LONG-TERM SUSTAINABILITY STRATEGY

Develop diversified revenue streams, including local services and a fractional ownership fund. Implement a community resilience plan against market fluctuations.

MEASURABLE COMMUNITY IMPACT

Define and track social, environmental, and financial KPIs. Reinvest generated value back into local infrastructure and education.

EMPIRE NEXUS

MAKING THE UNSEEN — UNDERSTOOD

www.ingramcontent.com/pod-product-compliance
Lightning Source LLC
LaVergne TN
LVHW050808061126
851857LV00012B/722